Transmediterranean

Diasporas, Histories, Geopolitical Spaces

P.I.E. Peter Lang

Bruxelles · Bern · Berlin · Frankfurt am Main · New York · Oxford · Wien

Joseph PUGLIESE (ed.)

Transmediterranean

Diasporas, Histories, Geopolitical Spaces

This book was made possible through the financial support of the Innovative Universities European Union and the Department of Media, Music and Cultural Studies at Macquarie University

Illustration on front cover © 'Capo Vaticano, Calabria,' by Joseph Pugliese

Éditions scientifiques internationales
Brussels, 2010
1 avenue Maurice, B-1050 Brussels, Belgium
info@peterlang.com; www.peterlang.com

Printed in Germany

ISBN 978-90-5201-619-1
D/2010/5678/37

Library of Congress Cataloging-in-Publication Data

Pugliese, Joseph, 1959- Transmediterranean : diaspora, histories, geopolitical spaces / Joseph Pugliese. p. cm. Includes bibliographical references and index. ISBN 978-90-5201-619-1
1. Mediterranean Region—Social conditions—1945- 2. Mediterranean Region—Emigration and immigration. 3. Regionalism—Mediterranean Region. 4. Geopolitics—Mediterranean Region. 5. Group identity—Mediterranean Region. 6. Transnationalism—Social aspects—Mediterranean Region. 7. Transnationalism—Political aspects—Mediterranean Region. I. Title. DE100.P84 2010 909'.098220825—dc22 2010018121

CIP also available from the British Library, GB

Bibliographic information published by "Die Deutsche Nationalbibliothek"
"Die Deutsche Nationalbibliothek" lists this publication in the "Deutsche Nationalbibliografie"; detailed bibliographic data is available on the Internet at <http://dnb.ddb.de>.

Contents

Acknowledgements

This book draws upon a series of papers delivered at a symposium, "Relational Dis/Locations: Mediterranean Cultures in Transnational and Translocal Contexts", held at Macquarie University, Sydney, Australia, 28 November 2008. The convening of this symposium was inspired by oral stories my late father, Carmelo Pugliese, told me about the village of our birth, Spilinga, Calabria. My father's oral narratives articulated histories of transmediterranean affiliations, genealogies and cultural exchanges between Calabria, the Middle East and North Africa that largely fail to register in official history books. The stories my father told me provoked me to write a paper, "White Historicide and the Returns of the Souths of the South" (Pugliese 2007), that attempted to document and piece together the shards of history otherwise subjugated and effaced by dominant Eurocentric histories. The symposium and this book are a continuation of this effort to name, identify and materialise the transmediterranean lines of connection that defy the borders and boundaries of nation-states and xenophobic nationalisms.

A number of people have been crucial in helping bring this book to fruition. Special thanks to Suvendrini Perera for encouraging me to pursue the publication of the symposium's papers into a book. My exchanges and dialogues with Suvendrini have shaped my intellectual life in manifold ways, and her own work on cultural junctions in the context of Australia and Asia has been a source of inspiration. I am particularly grateful to Blanche Menadier, Deputy Director of the Innovative Universities European Union Centre, who supported both the symposium and this book with an enthusiasm that was backed by invaluable financial funds. My thanks to the Innovative Universities European Union Centre for their financial support. I am grateful to Jon Seltin, who provided critical logistical planning and support that ensured the success of the symposium. My thanks also to the old Department of Critical and Cultural Studies and the new Department of Media, Music and Cultural Studies, Macquarie University for their support. Thank you to Emilie Menz at PIE Peter Lang for her crucial support of this project. This book would not have been possible without the enthusiasm and commitment of a transnational group of scholars that gave generously in realising a transmediterranean community in practice. Finally, particular thanks to Noah Bassil, Maria Giannacopoulos, Goldie Osuri and Ilaria Vanni for assisting me in the editing of this book.

The essay, "Transnational Carceral Archipelagos: Lampedusa and Christmas Island", was first published as "Crisis Heterotopias and Border Zones of the Dead", *Continuum*, Vol. 23, No. 5 (2009): 663-679, and has been reproduced with permission from Taylor & Francis. The first version of the essay "Imaging Italians Abroad. The 2008 Italian Political Election Campaign in Australia" was co-authored with Antonella Biscaro, whom we thank very much for her contribution.

INTRODUCTION

Transmediterranean Cultures in Transnational Contexts

Joseph PUGLIESE

Associate Professor at Macquarie University, Sydney

Diasporic Traversals

The Mediterranean Sea is both a physical space and a place of memory and affect inscribed with complex sedimentations of history, culture and politics. It has been a grand theatre that has witnessed the rise and fall of cultures, societies and empires. Yet, precisely because of its constitutive and agenda-setting historical role, it has too often been theorised as a body that is self-identical to itself – that is, the Mediterranean, in the discursive formations of received knowledge, has remained an entity conceptualised as a fixed geographical locus mapped and regulated by a series of longitudinal and latitudinal coordinates. The premise that underpins the essays collected in this book departs from this type of conceptualisation of the Mediterranean in order to ask the difficult question: how is the Mediterranean not identical to itself? This question, in other words, rejects conceptualisations of the Mediterranean perhaps best encapsulated by the imperial tag: *mare nostrum*. The Mediterranean, as *mare nostrum*/our sea, is already marked out as a transparently known entity inscribed by fields of force, possession and containment. Against this imperial and received conceptualisation of the Mediterranean, the essays in this book propose a transmediterranean traversed by lines of contestation, displacement and heterogeneous reconfigurations. There is no "essence" of Mediterranean culture or cultures as such, only dynamic, transformative and heterogeneous figurations that connect back to this geopolitical locus, even as they dis/locate and reinvent its histories, legacies and cultural affiliations.

Critically thinking and writing about the Mediterranean as an entity "outside itself" could perhaps only effectively be staged by subjects who are "outside themselves". In fact, all the scholars represented in this book are, in varying degrees, diasporic subjects who can trace lines of connection back to the Mediterranean, even as they continue to live and work well beyond its physical presence and cartographic parameters. As such, the essays in this book, precisely as they theorise and represent Mediterranean cultures in all of their multiplicities, are concerned principally with phenomena of dispersion, diaspora and the transitive movement encapsulated by the figure of the *trans*[1] – specifically, the *transmediterranean*. If the figure embodied by the trope of the *trans* signifies anything, it is the transversal movement that cuts across authorised borders, moving not vertically or horizontally but diagonally across space. The essays in this book, in varying modes, locate themselves on the fraught and unstable fault line of the border – as site to be traversed, problematised and contested. If the rhetorical force of *trans* resides in its ability to establish, through transversal crossings, systems of relation between otherwise disparate subjects, then its productive power must be seen as structurally generative of new and often seemingly untenable reconfigurations. Situated within this philosophical-tropological context, the figure of *trans* can be seen to be charged with the power to call into question the doxa of origins – of categories and identities that are at once epistemological and ontological, corporeal and geopolitical (see Stryker 2006). The figure of *trans*, through its constitutively tropological movement of transpositioning, reconfigures seemingly originary categories such as "homeland", "nation" and "identity" and, in the process, overturns the law of separation between such seemingly autonomous categories thereby disclosing their constitutive enmeshment and interdependence.

The figure of *trans*, in crossing seemingly autonomous and naturalised categories such as "nation" and "identity", brings into focus the iterable, always already constructed status of these categories. Inhabiting these categories from the inside and transfiguring them from dis/locations well beyond the physical parameters of the Mediterranean, the trope of the *trans* productively calls into question the "metaphysics of the proper" (Derrida 1976: 244) and the attendant border-control agents preoccupied with policing the propriety of the pure and the legitimate, the legal and the authorised in the face of border incursions. Furthermore, the power of the figure of *trans* resides in its ability to interrogate originary categories and to condition the field of inquiry simultaneously as it refuses – through the tropic force and animating logic of transpositional movement – to belong to the very borders and limits of the field that it traverses and unsettles. Animating this critical theorisation of

[1] This meditation on the figure of *trans* was first delivered at the Transsomatechnics Conference, Simon Fraser University, Vancouver, Canada, 2 May 2008. I am grateful to Susan Stryker for offering me this opportunity to think through the power of *trans*; her work in this area has been inspirational.

transmediterranean cultures is, of course, the landmark work of Paul Gilroy on transatlantic cultures and their politico-cultural ability to "transcend both the structures of the nation state and the constraints of ethnicity and national particularity" (1993: 19).

Rather than view the figure of *trans* – in all its multiplicity of bodies, forms and subjects – as a mere supplement, however dangerous, to established categories, essences and normative formations, what the essays in this collection evidence is a field of cultural dispersions and disseminations that ruptures established binaries such as native/immigrant, legal/illegal, authentic/reproduction and so on. So, for example, what would appear to be the quintessence of "native" Italian cuisine (dried pasta, canned tomatoes, and olive oil) is in fact the result of the formative influence of Italian immigrant communities in shaping economic and cultural practices back "home". In her critical study of the cultural politics of Italian food, Carol Helstosky (2004: 8) has mapped the dimensions of this native/immigrant transposition:

> These waves of [Italian] migration had both direct and indirect impact on the food available within Italy. Relatedly, immigrant remittances and the decrease in competition for jobs within Italy meant the living standard improved over the course of the early 20th century. And, as Italians began to leave the peninsula for greater economic opportunities in the Americas and elsewhere, they sought to recreate familiar dishes. This led to a growing body of consumers for Italian products (dried pasta, canned tomatoes, olive oil), which in turn greatly aided the development of certain food industries within Italy. Only after substantial numbers of Italians abroad began consuming these foods did domestic production furnish products for Italians at home. Ironically, it was because of the 'imagined communities' outside Italy that the food industry inside Italy produced the goods that became the foundations of Italian cuisine.

Transposed to this field of dispersions, I want to reflect on two things: on the dimensions of the transformative power of *trans* and, simultaneously, on those prescriptive categories and policed borders that defy this reconfiguring power of transpositioning, remaining largely untouched even after the tropic movement of crossing has transpired; and, finally, on the possibility of alliances between and across transmediterranean communities as a way of mobilising collectively to transform precisely what remains violently in place even after the always risky and vulnerable act of crossing.

As I observed above, the power of the figure of *trans* resides in its interrogative and transformative ability. I want, at this juncture, to begin to trace what remains recalcitrant to these transformative movements of transposition and reconfiguration. And I am thinking here of the manner, for example, in which undocumented immigrants, refugees and asylum seekers so often fall victim to the violence of militarised and policed national borders as they attempt their border crossings – leaving intact, despite their sacrificial deaths in the very movement of transnational crossing, the very lethal borders they attempt to evade, subvert or cross.

In a video installation titled "Made in Ghana (Man in a Box; On a Journey)", the contemporary Ghanaian artist Mawuli Afatsiawo depicts a man in a shipping container-like box.[2] Literally boxed in on all sides, the man is shown encased in his container, crawling on hands and knees, trying desperately to find an exit. This picture of imprisonment and entrapment in a shipping container is juxtaposed against a backdrop of bustling shops and markets in which the viewer witnesses the free and unrestricted flow of goods and commodities. In his video, Afatsiawo graphically materialises the stark disjunctions that inscribe globalisation: the free flow of commodities is predicated on the restrictive movement of the subjects of the Global South. In their attempt to circumvent this geopolitical double standard, refugees, asylum seekers and undocumented immigrants, clandestinely insert themselves within shipping containers, vans, trucks and other modes of transport instrumental in enabling the global flow of goods and services. These modes of transport become at once clandestine forms of mobility and suffocating containers of trauma and death for thousands of subjects from the Global South attempting to make their way to the countries of the Global North (see Pugliese 2009).

The nuanced and complex mappings of transmediterranean histories, cultures and geopolitical spaces that are staged by the essays in this volume affirm the transformative power of transversal crossings, even as they soberly mark the violent structures that remain in place despite this movement.

Transmediterranean Instantiations

In the opening essay of this volume, "Alien Conscription, Australian Sovereignty and the Vietnam War", Maria Giannacopoulos proceeds to map the complex lines of dislocation and affiliation between Greece and Australia. Situated within a theoretical frame concerned with issues of citizenship in times of war, Giannacopoulos unfolds a previously untold story that brings into focus the manner in which economies of exploitation are opportunistically mobilised by the nation-state in times of war. Beginning with the enforced deportation from Greece of millions of

2 Afatsiawo's video installation was screened at the 2006 Contemporary Commonwealth exhibition, Australian Centre for the Moving Image, Melbourne, February-May 2006.

subjects post-Second World War, she cites her father's eloquent observation: "We had to leave so the rest could live". As Giannacopoulos explains, her father and his family fled one imperially driven war only to be enmeshed within yet another war of empire and ideology: the Vietnam War. In the context of this war, Giannacopoulos' father is swept up in a bizarre and arbitrary national enlistment scheme that, through the exercise of white Australian sovereignty, disregarded the non-citizen, alien status of non-British immigrants in order to make them available to be conscripted into the war. That Giannacopoulos' father was spared this violent fate is only due to the fact that his number was literally not forthcoming in the draws of lottery-like ballots that, through biopolitical regimes of governmentality, transmuted lives into numbers. As she concludes, this biopolitical program of transmuting aliens into militarised citizens is continuing today under the auspices of the US war in Iraq and the targeting of the "Hispanic Market", with the promise that enlisted subjects can convert their alien status to that of citizen.

The privileges of citizenship are what are enunciated in the opening lines of Lara Palombo's "The Drawing of the Sovereign Line". Beginning from what is apparently a personal and circumscribed location, she narrates the story of her southern Italian grandmother's failure to access nursing home facilities in the northern Italian city of Como, despite having spent years of her productive working life in the region, because of the exclusionary and racialised laws of "Territorial Welfare". This instantiation of a seemingly individual and personal exclusion, however, is soon situated by Palombo within both national and transnational relations of power that are shown to be gendered, racialised and underpinned by neoliberal values. Following in the wake of Giannacopoulos' analysis of the operations of white sovereignty on Greek immigrant subjects, Palombo further qualifies this term by delineating the hetropatriarchal dimensions of transnational forms of white sovereignty as they operate upon Southerners (including southern Italian citizens and non-citizens from the Global South such as refugees, immigrants and the undocumented) within different geopolitical contexts. In her mapping of the complex operations of white sovereignty, Palombo brings into focus past and continuing colonial histories that cross and criss-cross both the Italian and Australian nation-states, marking, in the process, lines of connection and refraction between the two.

In her "Giving the Voiceless a Voice: Saving Third World Women Through a Western Lens", Seren Dalkiran stages a thorough critique of the paternalism of Eurocentric feminists who continue to frame Third World women as in need of "saving" – from themselves and their cultures. Unpacking the catalogue of stereotypes that continue to script non-European women from the Global South in contemporary Europe,

Dalkiran mounts her critique of the way in which Eurocentric fetish-objects such as the Muslim veil are mobilised by western feminists and others in order to homogenise all Muslim women as oppressed victims. Focusing on the unstated Eurocentric norms that underpin such stereotypes, she proposes other "lenses" through which to begin to view and understand such practices as veiling. As Dalkiran argues, outside the paternal and homogenising western lens, such practices emerge, for many Muslim women, as both liberatory and essential in establishing "social collective environments" of "shared standards, religious beliefs and moral values". Dalkiran productively develops her thesis by disclosing the complex political and cultural forces that the controversial former Dutch member of parliament, Ayaan Hirsi Ali, magnatised through her very public declaration on the "abuse of Muslim women and the crisis of multiculturalism in general".

Dalkiran's project of giving the "voiceless a voice" is reconfigured along entirely other, but related, lines by Ihab Shalbak in "Edward Said and the Palestinian Experience". In this eloquent meditation on the role of the public intellectual, Edward Said, in his giving voice to the Palestinian struggle, Shalbak traces the larger configurations of empire, colonisation, and the urgent issue of Palestinian self-determination within the context of this singular figure and his passionate commitment to a cultural politics of writing and literature. Beginning with a recount of a tragic story by the Palestinian writer, Ghassan Kanafi, Shalbak demonstrates how, for Said, the relation between politics and the cultural (*qua* literary) was indissociable, that one could not be understood without the other. Nothing exemplifies this premise with more force and conviction than this observation by Shalbak, once it is situated in the ethnogenocidal realities of contemporary Palestine and the cultural-literary compulsion to materialise Palestinian survival and existence in the face of violence and prospective decimation: "In the Palestinian case other worlds may not merely exist but they ought to exist out of necessity if not by choice". This was one of the fragile truths that Said strove to impart in the course of his politico-cultural life. Shalbak concludes his chapter by foregrounding the importance of Said's transitive approach in his politico-cultural work, arguing that this approach enables the materialisation of the historical nature of points of origin, simultaneously as it allows for the identification of the contemporary continuities that flow from such originary points.

It is precisely this transitive approach to the study of history and contemporary culture that Noah Raffoul Bassil stages in his "The Construction of the Colonial State in Sudan: Tribe, Region and Race and Colonial Power in Darfur". With incisive skill, Bassil proceeds to materialise the stratified history of colonialism and empire in the context of the

Sudan. He at once brings into focus the pre-colonial complexities of identity formation in Darfur and the manner in which imperial powers such as Britain, once they colonised the Sudan, proceeded to exploit, through the colonial rule of divide and conquer, pre-existing tribal and regional identities. In his tracking of the imposition of the so-called "thin white line" in the everyday colonial administration of the Sudan by the British, Bassil discloses the manner in which Eurocentric racial hierarchies functioned to classify, tribalise and, critically, to isolate groups that had previously co-existed together. The results of the administrative dimensions of this human geography, as Bassil demonstrates, was the colonial rupturing of previously existing transafrican and transmediterranean intercultural and economic flows and the consequent rigid boundary-patrolling of demarcated groups situated within distinctly unequal relations of power: "The benefits bestowed on tribes, including recognition of semi-autonomy and an allocation of land, or a tribal *dar*, left those groups that failed to be recognized as tribes in their own rights, as outsiders in a system dependent on tribal identity". In his rigorous historicising of the colonial and racialised history of Darfur, Bassil overturns the western stereotype of the innately genocidal African (as represented in western media by such conflicts as Rwanda and Darfur) by evidencing the colonial and imperial processes that have been instrumental in engendering conflict and war in the region.

In my essay, "Transnational Carceral Archipelagos: Lampedusa and Christmas Island", I proceed, like Bassil, to materialise histories of colonialism and biopower in the context of two islands, Lampedusa, off the coast of Sicily, and Christmas Island, off the coast of Western Australia. By examining the manner in which both these islands have been mobilised by Fortress Europe and Fortress Australia as frontline spaces that must thwart, through imprisonment and deportation, the landfall of asylum seekers and the undocumented on their shores, I attempt to mark their fault line status in the geopolitics of North/South relations. In the process, I also draw attention to transmediterranean historical and cultural flows that mark relations between Italy and Australia. Even as both Lampedusa and Christmas Island are sites marked by the harrowing presence of immigration detention prisons, they are also places that are destinations for luxury holidays. By drawing on Michel Foucault's concept of *crisis heterotopias*, as spaces that can simultaneously accommodate violent and contradictory differences (for example, vacation resorts and prisons, tourists and imprisoned refugees), I theorise the biopolitical relations that inscribe and organise such sites. Situating both these islands along the fault line of the border, I conclude by bringing into focus the border zones of the refugee dead that circumscribe both these islands.

The historical status of transmediterranean intellectual, racial and political flows between Italy and Australia is what Gaia Giuliani brings into focus in her "Whose Whiteness? Cultural Dis-Locations Between Italy and Australia". By drawing attention to the influential racial theories of the Italian criminal anthropologist Cesare Lombroso (1835-1909), Giuliani argues that such racial theories played a pivotal role in shaping Australian intellectual and political race thinking from the end of the 19th century to the close of the 1930s. She contextualises these influential racial theories within a medico-scientific field, addressing their material ramifications through a critical examination of the work of an Australian doctor, Raphael Cilento (1893-1985). Cilento, Giuliani argues, was an important figure in shaping both debates and policies centred on the colonial administration of the tropics in Australia and their attendant concerns about racial purity and hygiene. Giuliani proceeds to unravel a complex and intertwined Italian and Australian transmediterranean history underpinned by environmental, racial, medical, anthropological, eugenic and political theories that were instrumental in constructing transnational racialised ethnoscapes in both Europe and Australia.

The intertwined histories of transmediterranean cultural flows and exchanges are emblematically embodied by the Greek Australian writer and political activist, Alekos Doukas, in Petro Alexiou's "Alekos Doukas (1900-1962): A Dis-Located Life in the Shifting Terrain of the Eastern Mediterranean". In this essay, Alexiou stages a critical archaeology that refuses reductionist categories and homogenising labels. On the contrary, the transitive power of the figure of the *trans* (transmediterranean, transnational, translocal and so on) is here shown to breach such facile forms of containment: Greece, Turkey, the Mediterranean, the Ottoman Empire and Australia are all shown to be inextricably connected through the singular subject of Doukas. Alexiou unravels differential layers of historical, political and cultural sedimentation that shaped and were, in turn, reoriented and transformed by Doukas in the course of his diasporic life. As with so many of the other essays in this collection, fundamental questions on race, nation and colonialism are raised by Alexiou in the course of his study of Doukas. The importance of Alexiou's work lies in the manner in which an individual life-history becomes the vehicle through which to address complex transnational forces that usually fall outside of the concerns of traditional biographical studies. This enables him to flesh out a biographical life of Doukas that does not resile from naming contradictions, as well as celebrating multiplicities.

The identification and naming of a series of political contradictions is what Ilaria Vanni sets in motion in her "Imagining Italians Abroad:

The 2008 Italian Political Election Campaign in Australia". Through her delineation of a series of sardonic vignettes that figure the literal playing out of the recent Italian election campaign amongst Australia's Italian diaspora, Vanni effectively identifies "The ability of diasporic imagination to produce figurations that are both in conjunction and disjunction with the mythographies of the country of origin". Situating her study within the localised context of the nuanced and ephemeral practices of diasporic everyday life, Vanni demonstrates how the very ordinariness of such everyday diasporic practices "mobilise[s] a set of representations and self-representations that has little to do with the grandeur of Italian mainland political imagination". In effect, what Vanni so cannily discloses through this study are micro-political agentic practices that, through their deployment by transmediterranean diasporic subjects, are grounded in the realm of the senses, affect and the everyday and that contest, disperse and reconfigure nationalist political ideologies.

The transitive force of transmediterranean crossings in the context of nationalist fictions and mythographies is what is also critically examined in Goldie Osuri's "Transmediterranean Dispersals: Mazzini, Hindu Nationalism and Sonia Gandhi". Osuri's focus on a series of transmediterranean dispersals, that are embodied by such historically different figures as the Italian nationalist Giuseppe Mazzini (1805-1872) and the contemporary political figure Sonia Gandhi, enables her to set in train the productive unsettling of reductionist categories and self-identical formations. Consequently, the system of relations that holds between nation, religion and racialised embodiment is both unsettled and transfigured through Osuri's innovative disclosure of the links between Mazzinian theories of nation and religion and Sonia Maino Gandhi's incorporation and productive reorientation of such theories. In the face of homogenising and exclusionist Hindu nationalist ideologies, Sonia Gandhi emerges, in Osuri's words, as a trans-figure that contests "fictions of purity and homology – of blood, body, territory and nation". In her transitive study of Sonia Gandhi in the context of the politics of the Indian nation-state, Osuri underscores how these same nationalist fictions of purity are too often mobilised by various nationalist bodies not only to exclude otherised identities within particular geopolitical formations, but also in order to foment and legitimise violence against target subjects. In advocating transnational moves that question and contest such nationalist fictions, and in the context of a country, India, "whose metropolitan cities boast of road-side eateries which advertise curries, pizzas, and pastas on the same board, Indianness", Osuri concludes, "cannot be anchored to a homogeneous idea of culture or nation. This is the lesson that a mapping of transmediterranean dispersals has to offer us".

The "lesson" offered by Osuri perfectly encapsulates the key concerns of this book. If the study of transmediterranean cultures has anything to offer, it is precisely a lesson on how to contest homogenising and exclusionist categories that continue to reproduce both symbolic and physical forms of violence. The transmediterranean flows that are mapped in this book testify to embodied histories of relations, connections, exchanges and interdependencies that refuse self-identical categories, precisely as they open up the heterogeneous routes and heteronomous relays of transitive movements and their too often occluded histories. The essays in this collection are situated on a politico-historical fault line that is inscribed by globalising calls for economic liberalisation and the demolition of trade borders and, simultaneously, by the passing of ever more draconian nationalist policies and exclusionary racist laws delimiting and criminalising the movement of people from the Global South to the Global North. Situated in this charged context, the essays in this collection articulate both the violent legacies of previous racist and colonialist histories, whilst also uncovering previously effaced transmediterranean flows and exchanges that breach and productively reconfigure bodies and nations that claim to be self-identical to themselves.

Alien Conscription, Australian Sovereignty and the Vietnam War

Maria GIANNACOPOULOS

Affiliated Fellow, PhD, at Macquarie University, Sydney

Introduction

When my father casually mentioned at Christmas lunch a few years back that he *almost* had to serve in Vietnam for Australia as a young migrant, the 'almost' aspect of his story could have made it a non-issue. But what triggered my interest was that it seemed strange that a migrant or *alien*, as the official discourses of the time named people like father, would be required to serve for a country to which he had no official claim, in which his status was still provisional. This unofficial and *almost* story, then, becomes the pretext through which the legitimacy of claims relating to the state's ability via law to engender racial equality may be judged. This essay is set both in the past and present, it engages with the question of conscription during the Vietnam War, it takes up the question of Australian values and citizenship both during that time but also in the contemporary climate where being Mediterranean persists as an issue to be managed by the white Australian state. My father's *almost* story of Vietnam and alien conscription becomes a vehicle through which an otherwise buried aspect of Australia's history can come to light. It presents an opportunity to open up an area of scholarship that is not only relevant for historical reasons but can also provide a critical mode through which the racial supremacy of the Australian state may be challenged. The story of *alien* conscription told here is the story that sees the multiple modes of imperial power and control collide: from poverty stricken Greece under British and then US control to an Australian state born of British imperial violence which maintains its racial supremacy precisely through manoeuvres such as the conscription of migrant youth. Such enactments of white Australian sovereignty are precisely where the colonial violence of the Australian state are both obfuscated and naturalised.

From this time forward, under God
I pledge my loyalty to Australia and its people
Whose democratic beliefs I share,
Whose rights and liberties I respect, and
Whose laws I will uphold and obey.[1]

Dad' Story

The official guide to citizenship tells the prospective citizen that with these words, new citizens become part of the Australian community. They promise loyalty, respect and obedience to the law. They share freedoms, responsibilities and privileges (Commonwealth of Australia, 2007: 3). The pledge of citizenship is seen to have the power of transmutation since it can convert *aliens* into Australian citizens. But this official document does not assist in comprehending why my father in 1967 was required to register for military service during the Vietnam War. This was required of all Australian youth upon turning 20 but my father, although youthful, could not yet be classified as Australian. His transmutation into an Australian via citizenship did not occur until February 5, 1975. But in April 1966, cabinet endorsed the principle that aliens who had chosen Australia as their home should be eligible for national service, a change which came into effect in 1967 (Edwards, 1997: 103; Jordens, 1997: 175). Contrary to the assumptions on the use of the term *alien*, my father's place of origin was not outer space and unknown. He had migrated from Himadio, a rural village of Greece not far from the site of ancient Olympia in the Peloponnese, aged 17 in 1964. His migratory journey was a choice made under coercion, the type of coercion that was born of poverty in a place that had not been able to recover from the effects of older wars.

The Greece that my father was born into was not the kind of place that can be described as having afforded its people the degree of choice that is presumed in the policy statement set out by the Australian Parliament of 1966 when they suggest that the choice to migrate, since citizenship was absent, was sufficient to qualify young men like my father for war. The economic problems that Greece faced in the first part of the 20th century were greatly exacerbated by the Second World War. As was the common practice of the Axis occupiers, they levied "occupation costs and other exactions on a scale which not only denuded the country of its meagre resources but threatened to plunge it into bankruptcy" (Papandreou, 1971: 15). Papandreou reports that so serious was the situation that even Mussolini was moved to plead the cause of Greeks to Hitler (1971: 15). Nazi Herman Goering's response was one

1 Australian Citizenship Pledge, Commonwealth of Australia 2007.

of indifference, "We cannot worry too much about starvation among the Greeks… perhaps it is well that it should be so, for certain nations must be decimated. But even if it were not so, nothing could be done about it" (quoted in Papandreou, 1971: 15). The end of the Second World War signalled the beginning of the civil war between the British sponsored regency of the Archbishop of Athens and the Communist guerrilla bands in the mountains of the north. In March 1947 the British Government announced that it could no longer support the regency which it had placed in power and would abandon the country to the Communists (Papandreou 1971: 16). It was at this point that a US imperial presence, hastening to act at the thought of the country being left to the Communists, came to immediately impact the lives of those already impoverished:

> Under the Truman doctrine, which marked the first American experiment in political intervention outside of the New World, some 300 million dollars was poured into Greece, most of it in military aid. Along with the dollars came 'military advisers,' to instruct the Greek Army in guerrilla tactics. Under the impact of the American intervention, the Communists by 1949 had suffered severe setbacks; and by 1950 the Communist Party had been outlawed in Greece. Constitutional government – imposed by virtue of foreign intervention – was triumphant, and Greece, we were told, had been made safe for democracy (Papandreou, 1971: 16).

In the name of constitutional government and democracy then, the population continued to endure major economic hardship since the US funds were used to serve imperialist agendas rather than for the alleviation of suffering of the population. Papandreou reports that these funds were directed to immediate relief rather than to long term reconstruction and development (1971: 16). But the focus on immediate relief although logical that it should be prioritised, simultaneously functioned to curb the threat that dire poverty posed to capitalism's allure than to the hardship endured by the people. In 1949 approximately 700,000 people, nearly 10 percent of the population, were homeless refugees awaiting resettlement in their devastated villages (Papandreou, 1971: 16). Almost a third of the population were wholly or partly dependent on the state for their day to day survival. The overall cumulative effect of the world war and the civil war was, at the beginning of the 1950s, the complete dislocation of the national economy of the country (Papandreou, 1971: 16). In 1961 seasonal agricultural unemployment fluctuated between 10 and 25 percent depending on the region (Papandreou, 1971: 18). My father, from a large and poor rural family, at age 12 rode a pushbike for an hour and a half to reach the nearest city of Pyrgos to work as an assistant at the abattoirs. His father had been sent to fight in the Greco-Turkish war of 1919-22 in Asia Minor where he sustained injuries preventing his proper functioning in the agricultural context to which he had to return and work. There was never any compensation paid by the Greek state for injuries sustained because they said his case lacked the evidence required to prove that his injuries had been sustained during the war. Thus there was no such thing as a childhood for my father and his siblings. *We had no bread to eat* is the best English translation I can offer of how he describes that time. His life as a worker began long before he set foot on Australian soil and long before he became an adult.

Under those circumstances, emigration contributed towards a short term, partial solution of the problem of unemployment. As Greece emerged into the 1960s the loss of population through emigration at first equalled, and then surpassed, the natural growth of the population (Papandreou, 1971: 18). *We had to leave so the rest could live* is how my father describes it. This was possible because in 1945 Australia was concerned with increasing its population and developing its industry whilst simultaneously upholding the colonial ideal of a white Australia (Alexakis and Janiszewski, 1998: 17). So when the migration agreement was signed between Greece and Australia in 1952 through the Intergovernmental Committee for European Migration (Alexakis and Janiszewski, 1998: 17), Greeks from the poorest regions of the country became the very currency with which this mutually beneficial bargain could be made. For the Greek state the benefit was to de-populate and to export its poverty and for the Australian state, a provisionally white body could service its desire to populate, industrialise and continue to colonise. So as a young boy of 17, my father was sent to Australia. He was sent by his family, but also by the accumulated and devastating effects of war and imperial control first of Britain and then of the US. These imperial interventions in the name of democracy and economic recovery paralysed the country to the point where families had to sacrifice some members of their family for the sake of survival. This choice, to migrate, should not therefore be mistaken with the choice that is assumed and supposed by those who positioned and continue to position migrants as willing participants in the workings of white Australian sovereignty.

Conscription of Aliens

The *National Service Act 1964* was passed on November 24 and it required 20 year old males to serve in the Army for a period of 24 months of continuous service followed by three years in the Reserve. In 1965 the *Defence Act* was amended to provide that conscripts could be obliged to serve overseas and in March 1966 Prime Minister Harold Holt announced that national servicemen would be sent to Vietnam in units of the Australian regular Army (National Archives of Australia,

Fact Sheet 164). Harold Holt, who had just taken up the Prime Ministership, continued Robert Menzies' desire to commit Australian troops to Vietnam in support of the US citing aggressive Communist developments in Asia as the pretext. He said:

> Our first objective is to help the Government of the Republic of Vietnam, at its request and in light of our own assessment of the nature of the conflict in Vietnam, to resist the armed aggression of Communist North Vietnam from the threat of oppression and terror... The third objective is to leave no one in doubt that we in Australia are prepared and resolved to honour our treaty commitments and our alliances... we will ensure that the spread of Communism in South East Asia is checked (quoted in Edwards, 1997: 101-102).

Between 1964 and 1972, 804,286 twenty year olds registered for National Service (Langford, 1997: 369). Whilst registration was compulsory, the final decision as to who would go to war was decided by ballot. Two ballots were conducted each year. The ballots selected several dates in the selected period and all males with corresponding birthdays were called up for national service. The ballot was conducted using a lottery barrel and marbles representing birthdays (Langford, 1997: 374).

The US making the world safe from Communism and therefore for democracy would once again impact on my father albeit in a different continent far from the Mediterranean and it would most definitely impact on the 63,000 youth who were conscripted and the 19,000 who served in Vietnam (National Archives of Australia, Fact sheet 164). At first, non-naturalised immigrants living in Australia who were not from the UK were not required to register when the national scheme was introduced, in keeping with Regular Army regulations which did not allow non-British immigrants to enlist unless they intended to be naturalised (Langford, 1997: 367). The waiving of the non-British immigrant's obligation for national service came under criticism from the public, the press and Parliament on the grounds that immigrants living in Australia should have the same responsibilities as Australian nationals (Langford, 1997: 367). The 1965 Citizenship Convention recommended that all migrants who had been in Australia for over 10 years should be automatically recognised as citizens. It was believed that since migrants "enjoyed social services" they should "accept the responsibilities of citizenship as well" (Jordens, 1997: 185). According to Langford, "after considerable debate and close examination, particularly over whether the call up of immigrants was consistent with international law, Cabinet endorsed the proposal that non-British subjects ordinarily resident in Australia were liable for national service" (1997: 367).

The decision made in April of 1966 was not announced till August of that year giving the Government four months to consult with relevant foreign governments. The Italian Government formally protested (Langford, 1997: 367). A representative of the Government of Malta visiting Australia publicly supported the decision of the Australian government (Langford, 1997: 367). Whilst consultation time with foreign governments was allowed it seems highly unlikely that in Greece, on the eve of a military junta, any serious attention was being paid to what was happening to the problem it had recently exported. In July of 1965 the King of Greece summoned Prime Minister George Papandreou and obliged him to resign. Immediately following this the King swore in a new Prime Minister. The armed forces were placed on alert, to crush the uprising of the centre and of the left that was said to be imminent ('Athenian', 1972: 53). The right was said to be overjoyed at this development as was the US embassy which appeared to have encouraged the royal initiative. The legality of this situation was in question but apologists for the King's intervention argued that he adhered to the letter of the amended 1952 Constitution according to which the King appoints and dismisses ministers ('Athenian', 1972: 53-54). Meanwhile representatives of the Greek and Italian immigrant communities living under the ethically suspect and highly mutable Australian legal system publicly criticised the decision. The Hansard of the House of Representatives dated August 31, 1966 reflects this:

> Mr Daly also presented a petition from certain Greeks migrants living in the Commonwealth opposing any proposition to conscript migrant youth for the Armed Services and the Commonwealth Government's decision to conscript youth for service beyond the Commonwealth and its territories, particularly in Vietnam, and praying that the Commonwealth Government reverse its present policy.

The record spartanly states that the petition was received and read. Peter Edwards' book *A Nation at War: Australian Politics, Society and Diplomacy during the Vietnam War 1965-1975*, was written as part of a longer project in collaboration with the Australian War Memorial and now circulates as the official history of Australia's involvement in Southeast Asian conflicts. In this official text the position stated is that:

> Those who objected were clearly more concerned with the prospect of fighting in Vietnam than with the general principle of conscription of unnaturalised migrants. With the exception of *The Australian*, major newspapers endorsed the policy as eminently fair, *The Age* commending the Government's adroit handling of a complex problem. The Government had not only defused a potentially explosive issue but had even won praise for firmness, fairness and sensitivity to public opinion (Edwards, 1997: 103).

The decision came into force in January 1967 and was in line with the Government's aim to treat all immigrants, whether British subjects or not, along the same lines (Langford, 1997: 367). Yet according to

Langford, immigrants were treated differently from naturalised Australians in that they were not called up until they had been in Australia for at least two years. Non-British subjects were not called up for service until they had reached the age of 21, to enable the individual as an adult to make the decision whether to remain in Australia and, if so, to apply for naturalisation and be liable for national service (1997: 367). Non-British migrants were permitted to leave Australia rather than be called up. In February of 1967 the *Nationality and Citizenship Bill* was passed in Parliament giving non British immigrants liable for national service the right of citizenship after three months of national service on the condition that they were efficient soldiers. This Bill was to ensure that no national serviceman would serve overseas before having the opportunity of becoming an Australian citizen (Langford, 1997: 367).

Citizens Lotteries and Alien Choices

The conscription of aliens has received very little attention as a worthy site of critical reflection and analysis. This can be gauged not only by the lack of work published around this occurrence but also from the modes of description adopted by those who have reported it. When official historian Edwards declares that those objectors to the Government policy were more worried about fighting in Vietnam than with the principle of alien conscription he sets up these categories as mutually exclusive with the implication being that those affected by both the principle and the practices of alien conscription were beyond comprehending the problematic and violent nature of the principle. The irony here is that as he positions those affected as unable to comprehend the principle or as being disinterested in it, he is engaging by way of this reporting in a practice that denigrates the principle as a site worthy of critique. Edwards raises the issue as an issue but then proceeds dismissively to diminish its importance by charging those affected, by implication, with traits of cowardice and disloyalty.

My father's birthday, August 6, 1947, did not get drawn from the ballot. August 7, 8, 9, 11, 12 did get drawn (Langford, 1997: 374). It was sheer chance that did not place him in the position of having to make an alien choice. Should he fight and possibly die for a strange country or be expelled to the country that had effectively expelled him? The right to decide to return to Greece is presented as a freedom, "non-British migrants were permitted to leave Australia rather than be called up" (Langford, 1997: 367). Both my parents insist that it was generally understood that if that *choice* of return was made one could never return to Australia. This is not choice. It is logically inconsistent that one can *choose* to be deported since deportation is a denial of liberty aimed at putting one back in one's place. As of April 1967 Greece had experienced a *coup d'état* and was to remain a military dictatorship until 1974 ('Athenian', 1972: 61-70). These were the choices. The country of residence, Australia, the prospect of Vietnam and the unrecognisable militaristic homeland which all looked gloomy and inhospitable; with Australia's violence being the least obvious since its manifestation came in the form of a lottery ball which when set in motion could dictate the fate of a young migrant under the guise of fairness, equality and luck.

Why, as Edwards reports but does not explore, is the policy of alien conscription "complex" and "potentially explosive"? What amounted to the "adroit handling" of it? Also silent on the complexities and controversies of such questions are Alexakis and Janiszewski who claim that their 1998 book and accompanying project *In their Own Image: Greek Australians*

> has contributed to the documentation of both the historical and contemporary Greek-Australian presence in this country… over the past sixteen years, we have gathered one of the most nationally comprehensive visual, oral and literary archives on Greek-Australians… The book's principal concern is to attempt to broaden public understanding of Greek-Australians by breaking down the alienating and simplistic stereotypes and misconceptions (1998: 8).

Alexakis's and Janiszewski's project highlights, among other important things the *Immigration Restriction Act of 1901* which made racial exclusion legal as well as major anti-Greek riots in Sydney in 1915 and in Perth, Kalgoorlie and Boulder, Western Australia in 1916 (1998: 16-17). The riots had been preceded by strikes and demonstrations against enemy aliens. Those protesting insisted on the internment of the 'enemy' aliens living within the nation (1998: 17). Further they report that in 1934, during the Great Depression, race riots again erupted in Kalgoorlie, with Greeks and other Southern Europeans the objects of racial hatred (1998: 17). During Second World War, Greek shopfronts in a variety of locations nationwide were smashed by those who mistakenly thought that the business enterprises were Italian owned (1998: 17). It is because they have reported such facts relating to explicit racism both emanating from the state as well as from the general population that I am critical of their simplistic reporting of the issue of alien conscription. They do not in fact term it *alien* conscription. In a short paragraph entitled "Military Involvement" they simply state that "Greek-Australians also undertook military service during the Vietnam conflict" (1998: 16). In their introduction they speak of their desire to break down "alienating and simplistic stereotypes" but ironically this does not include being critical of the way in which, at various points in white Australian history, the alien was an official category from which violent effects continue to flow, including *alienating* stereotypes. In fact, their descriptive statement relating to Greek Australian involvement in the Vietnam War is not entirely correct. This is because it describes Greek Australians as having participated in the war without accounting for the Greeks who were not yet Australian, that is, they were *aliens*, and yet were still compelled to fight for Australia. I am not suggesting that their desire to break down stereotypes is not important, rather I am arguing that to focus on stereotypes alone, even if that is to reverse them by showing the many great achievements of Greek Australians, cannot critically engage with the colonial operations of white power which have not ceased since the transformation of Greeks from aliens to citizens or since the lines of inclusion and exclusion have become less distinct. In 1974 when my father applied for citizenship and eventually pledged loyalty in early 1975, he was reluctant to do so. He was finally convinced by my mother who had just given birth to their second child. An older Greek friend, who had become a citizen and was professionally employed, scolded my father for wanting to remain Greek and for not wanting to become Australian, since this would keep him a *nobody* in Australia. But his reluctance was justified, for to pledge allegiance to the state and to Australian law seemed redundant as these things, in the absence of citizenship, had already been demanded of him. Jordens in her book *Alien to Citizen*, reports on this relation of coercion as though it was of no real significance: "The introduction of conscription for aliens in 1967 during the Vietnam War conferred on them what was generally perceived to be one of the major responsibilities of citizenship, whether or not they wanted to become naturalised" (1997: 175, emphasis added).

The issue of alien conscription as Edwards suggests was "complex" and "potentially explosive" precisely because of the histories that Alexakis and Janiszewski have identified. If Greeks were understood to be enemy aliens, how could they be legitimate soldiers of the Australian state? If citizenship is that which brings about rights and responsibilities on the part of the citizen (Australian Government, 2007: 3), then in its absence from where does the requirement to defend Australia and to respect and uphold Australian values derive? The answer provided to these questions in 1967 was also the reason that the policy is alleged to have been propelled: equality. Langford writes that

> the waiving of non-British immigrants' obligation for national service came under criticism from the public, the press and Parliament on the grounds that immigrants living in Australia should have the same responsibilities as Australian nationals... The decision... was in keeping with the Government's aim to treat all immigrants, whether British subjects or not, along the same lines (1997: 367).

So despite the anti-foreign sentiment expressed in a myriad of ways since the establishment of the colonial state of Australia, suddenly both population and parliamentarians were demanding equality and those in power made the legislative changes to reflect this. I suggest that the meaning of equality in this context was equality of responsibility without the equality of the privileges that might accrue to a citizen. If a non-citizen can be compelled to go to war then it is not citizenship that decides how a migrant lives and potentially dies but rather the laws of white sovereignty. Whilst the discourse of sameness and equality propelled the laws of alien conscription, equality in those circumstances could not amount to justice since the state obfuscated its white legacy to assign *alien* status in order to discriminate in favour of itself. Equality in this instance did not stretch to the possibility of a British subject facing deportation without the right to return for not going to war. Since the white subject was already deemed to rightfully belong in Australia, the logic of white sovereign rule would make this unthinkable. A contemporary Australian value, one which citizens must abide by upon taking the pledge of citizenship, is equality under the law. This means that nobody should be treated differently from anybody else because of their race, ethnicity or country of origin (Australian Government, 2007: 6). This statement conceals the violent potential contained within Australian law since it is the law that defines equality in ways that produce white sovereign rule as legitimate and just. As my father's story attests, even though citizenship demands allegiance to Australian law, Australian law is under no reciprocal obligation to confer the rights of citizenship before it applies its force.

Tony Birch has argued that Indigenous peoples "have for too long now been regarded as dispossessed in our own land, aliens in our own land, 'citizens without rights' in our own land" (2000: 3). This powerful assertion highlights in a different but related sense the way in which the category of the alien plays a central role in the operations of the white sovereign state. The alien does not exist prior to the exercise of white racial violence but rather is the effect of it, even if this category impacts on different peoples in distinct ways. The migrant alien did not exist prior to entry but was rather produced as such at the level of law upon entry to Australia. The Indigenous alien following on from Birch's statement, that is, the peoples whose land is stolen and become subjected to colonial power, is what founds that white law and therefore its power to designate others seeking to arrive or stay in the country as aliens. On this argument, Australia is founded and defended as a white sovereign state on multiple levels: through the alienation of Indigenous peoples from their sovereign status through dispossession, by sending those expendable migrant aliens to war to assert and defend Australia as a white sovereign state, a war which produced more refugee aliens for Australian law to contend with.

But there is another sense in which Birch's insight manifests, especially in regards to his description of Indigenous peoples as being "citizens without rights" since the introduction of conscription drew attention to this very issue. The *Commonwealth National Service Act*, which regulated conscription in the 1960s, specifically excluded Indigenous peoples from conscription (Jordens, 1997: 15). There were however conservative sectors of the community, such as the Returned Services League, the Country Women's Association and sections of the Country party who argued that since Aboriginal people enjoyed the rights of citizenship they should be required to fulfil its duties and be subject to conscription (Jordens, 1997: 16). This discourse of equality which masked violent inequalities mirrored the logic that led to the law being changed to conscript migrant youth. Although Indigenous peoples could not formally be considered aliens, their status of being alienated from their lands, that is to say the land deemed *terra nullius* and available for the operations of white law, provided the foundation for these violent debates and policies that designated who could be allowed meaningful entry to the category of citizen and who had to live and who had to potentially die to shore up the space of the white nation.

The Force of Australian Law

In order to critique the treatment meted out to those it designates as aliens, the immigrant status of white law must be brought to the fore. Australian law, at each historical juncture, has been continually engaged in the colonially violent dual action of erasing its migrant status in order to appear to rule with legitimacy. This legitimacy is the result of the colonial erasure of Indigenous laws which clears the legal landscape of other laws thus allowing white law to occupy the position of the law. One case in which this is manifest is the 1992 Australian High Court decision of *Mabo v Commonwealth (Cth)* where the legal doctrine that is said to have been foundational to the formation of the Australian state, *terra nullius* was overturned. In overturning this doctrine which describes the land taken by colonists as belonging to no one, white sovereignty was guarded as non-justiciable (Giannacopoulos, 2007: 48) and with this continued the project of producing the coloniser, its key people and key institutions, as non-immigrant. Aileen Moreton-Robinson has tracked the way in which the British Empire imposed itself through both colonisation and waves of migrants from Britain (2003: 24). She writes that "the British claimed the land under the legal fiction of Terra Nullius – and belonging to no one – and systematically dispossessed, murdered, raped and incarcerated the original owners on cattle stations, missions and reserves" (2003: 25). This point makes visible and explicit the direct link between migration and the legally produced doctrine that effectively cleared the land for filling. Moreton-Robinson's critical point is that "through the fiction of Terra Nullius the migrant has come to claim the right to live in our land. This is one of the fundamental benefits white British migrants derived from dispossession" (2003: 25).

Whilst Moreton-Robinson writes of British colonisation along with British migration as the foundational events in the illegitimate assertion of white sovereignty, Goldie Osuri and Bobby Banerjee contend that white diasporic events are rendered invisible by a series of manoeuvres including the strategic employment of the term *settler* (2004: 158). They draw upon the work of Mahmood Mamdani who states that "settlers are made by conquest, not just immigration. Settlers are kept by a form of the state that makes a distinction particularly juridical – between conquerors and conquered, settlers and natives" (quoted in Osuri and Banerjee, 2004: 158-159). Whilst Mamdani has written in the context of the African settler states, it has been argued by Ahluwalia that Mamdani's work is relevant to Indigenous dispossession in Australia since "the myth of *terra nullius* was dependent upon the non-recognition of the local population and the 'indigenisation' of their white conquerors" (quoted in Osuri and Banerjee, 2004: 159). The question of alien conscription animates Mamdani's claim: that the state *keeps settlers* by drawing distinctions, *particularly juridical* distinctions between categories of persons. Settlers and British migrants are not historical novelties as far as the operations of racialised power are concerned. They are generated by a law that obfuscates its status as immigrant, diasporic and racially violent. They are generated by a law that is indissociable from the operations of white sovereignty (Giannacopoulos, 2006). As such, the *non-migrant, non-diasporic* white sovereign can keep itself as mystically un-ethnic and transcendental by asserting its jurisdiction over the bodies it looks out toward.

Kafka's parable, *Before the Law*, is the story of an unnamed man whose entire life is spent begging and waiting for admittance to law. The law's doorkeeper never admits him and while at the front line of law's border-policing the man dies waiting:

> Before the law stands a doorkeeper. To this doorkeeper there comes a man from the country and asks to be admitted to the law. But the doorkeeper says that he cannot at present grant him admittance. The man considers, and then asks whether that means he may be admitted later on. 'It is possible' says the doorkeeper, 'but not at present'… The man from the country has not expected such difficulties; the law, he thinks, should be accessible to everyone at all times; as he now takes a closer look at the doorkeeper in his fur coat, at his large pointed nose, his long, sparse, black tartar beard, he decides that it is better, after all, to wait until he receives permission to enter. The doorkeeper gives him a stool and lets him sit down to one side of the door. There he sits for days and years (Kafka, 1992: 165).

Kafka's story is about the reign of law, the absolute rule of law, of *nomos basileus* that is of *law as sovereign.* Law needs *keepers to keep* it according to the story. Who are law's keepers? What is law? And who or what is external to law? The man from the country is prevented entry but is he outside of the reign of law? If his whole life is spent waiting, begging, complying, in the hope that the law will exercise its discretion and grant him entry, then I contend that the space he occupies is already *of* law in that it is a highly regulated space that demands compliance up until the point of death. Law in Kafka's parable is the place where law is *in force* even when, and perhaps because, subjects are not formally within its jurisdiction. In the same way that Kafka's law is unmarked so too does white law generate itself as ethnically neutral and as though its jurisdiction is without limits (see Derrida, 1992).

The power to conscript aliens is but one instance where an overarching white, non-ethnic, non-immigrant universalised notion of law can be kept and further entrenched. Irene Watson has written of the ways in which the "coloniser perceived this Nunga place as available to be filled with their 'beginnings' of history and 'evolving spirit'" (2002: 254). The white sovereign state's power of discretion to decide which persons will be deemed Australian and under what conditions should be seen as evidence of a still 'evolving' white Australian legal system that is proceeding *as though* the land, Indigenous land, is still empty and needing to be filled with the law of the coloniser. But this *evolution* requires that the coloniser's status be renewed, as a neutral identity at law, in order for colonial work to continue. The *outsiders* and *permanent immigrants* have instilled themselves as sovereigns who rule with law with such violent force, that they can literally write themselves out of the narrative even as they continue to colonise armed with this master script.

From Dispensable to Deportable Ethnics

It is because Australia still functions as though it can be filled and defined at the will and discretion of the guardians of white sovereignty, that non-British migrants become the middlemen for contemporary colonisation. When white law acts sovereign, hovers mystically above all other forms of law and sovereignty and is securely kept in place by a violent colonial law, then it is law that creates the national space of *Australia* and in so doing makes the nation intelligible and operable as a "white possession" (see Moreton-Robinson, 2004). Former Prime Minister John Howard in an address at the Hellenic Club in September 2006 said that Greeks "are 'a wonderful example' of integration" (ABC News, 2006). Surely I should have been happy with such a validation of the community I belong to. Instead these comments deeply disturbed me. How could an act of praise still be an exercise in racial violence?

Former Prime Minister Howard said that the Greek community has demonstrated how to embrace Australian culture. He said "The Greeks are just a wonderful example of how you do it, you integrate fully, you become part of the mainstream… Your first loyalty is to Australia, but that doesn't mean that you don't have a place in your heart for your home culture, and that's how we want it" (ABC, News 2006).

Here Howard was itemising the heart as the correct holding topos for love of Greece. The 'integrated Australian' must contain within the borders of his/her body those emotive aspects that could be threatening to the rational domain of loyalty to Australia. Howard was evoking the familiar rhetoric of the periods of integration and assimilation that defined Australia as white. The terms he employs, "integrate fully" and "loyalty" can only be intelligible if the assumptions that drive them have become so familiar that they cease to be visible. That Howard has the right to employ language which simultaneously assumes and asserts his right to decide and declare which *outsiders* and *immigrants*, apart from him and his kind, have successfully *integrated* is an exercise of sovereign power that continues to be produced at the level of law. When Howard exercises the logic of patriarchal white sovereignty (see Moreton-Robinson, 2004) he colonises in multiple ways: he produces ethnicity as something that can only be present in non-British people and their descendants; he acts to protect the white nation as produced by white law; and by silently drawing upon legal knowledge that has come to appear familiar and as common sense he colonises the space of what counts as legitimate knowledge about the nation.

Howard praises Greeks because they have become part of the mainstream. He suggests that the Greeks have hit upon the magic formula for becoming good *ethnic* Australians by having worked out the correct balance between loyalty to Australia and a place in one's heart for his/her home culture. Howard says "this is the way we want it" (ABC, News 2006). So according to Howard's two part formula, there are two homes. There is *our home* and there is *their home*. From this breakdown I draw the following equation: *they* who are still "they" even though *they* have "fully integrated" must be loyal to *our* home without it ever quite being *their* home. This is the way "we" want it says Howard. Who are the "we" in his formulation? If Howard's statement is taken seriously then my father is one of Howard's celebrated subjects. If my father has become part of the mainstream, if he has "integrated" fully but still pines, in the correct and measured amounts, for his homeland then is he part of the "we" who now validate Greeks? Is he part of the "we" that is patting Greeks on the back? Is dad patting himself on the back? After all it seems that he is the "we" and the "they" in this instance. Or is it that dad will only ever be capable of being patted on the back? I contend that Howard's white "we" excludes my father and not coincidentally has the same traits as white Australian law. It is a "we" that operates as racially unmarked and as being without ethnicity since Howard's "we" and white law are the epistemological centres from where the terms of these debates are produced and around which all discussions then pivot.

Howard made these comments against the backdrop of debates about whether new migrants will be required to undertake citizenship and "values" tests (see DIMA, 2006). Without entering into the discussion of what would constitute *Australian values* under this regime, it is sufficient to say that at that historical juncture, Howard was drawing Greeks under the umbrella of whiteness in order to perpetrate racial violence against other communities in Australia. He did this by erasing the struggles, past and present, that Greeks have experienced in Australia and by actively forgetting that they once did not qualify as white enough or worthy enough to enter Australia. And in fact when they did enter at different historical junctures, their presence was accompanied by an extensive array of racial violence, the proof of which resides in personal narratives, most of which have yet to be told. The officially homogenised category of "Southern Europeans" was considered far less capable of assimilation and so was historically far less desired than the British or Northern Europeans (Lopez, 2000: 43). It was in 1964, the same year of my father's move to Australia that the Department of Immigration's Assimilation Branch changed its name to the Integration Branch (Lopez, 2000: 62). The official workings of integration continued the logic of assimilation, with an added emphasis on the idea of "core values". Lopez states that "those who spoke of core values implied that the greater proportion of migrant cultural practices and national loyalties, considered not to transgress core values, could be tolerated" (Lopez, 2000: 58). When Howard's comments are placed against this historical backdrop, not only is it abundantly clear that in contemporary racial politics there is a unashamed return to the period of which Lopez writes, but also that Howard was using the same logic to praise Greeks as was used to punish them at other historical moments. The *familiar* and legally generated logic of integration re-enters popular debates whilst simultaneously never having been absent from them.

Howard's celebratory words of integration, uttered in 2006 were informed by the same discourses that in 1966 demanded *equality* of conscription for *alien* Greeks. His homogenising logic functions to place Greeks in opposition to other *others*. So in fact the validation that Howard offers can only be enjoyed if I/we ignore that this validation is in fact a form of colonial violence. Accepting praise from Howard means accepting, as valid, the white sovereign power from which that praise emanates as well as to sanction punishment of other communities who have not yet shown the compliance he says my people have shown. This benevolent *recognition* of Greeks as *well integrated* is more a strategic positioning in relation to white power than it can be a statement of fact. In fact, it is a gesture that is in line with ongoing colonial logic.

From Conscription to Recruitment

The practices of *alien* conscription which I have tracked in this essay have a contemporary counterpart in the current practices of the US Army, specifically in relation to its aggressive recruitment of Latinos/Latinas, particularly Mexicans to fight for the US in the war in Iraq. This is a context in which citizenship as social currency is being played out in the most dramatic terms. Jorge Mariscal writes that "During the early days of the US invasion of Iraq, the number of Spanish surnamed soldiers and marines killed or missing in action became evident. Latino youth were being disproportionately sacrificed once again, as in Vietnam" (2003). He also writes that many of the Latinos/Latinas in the armed forces are non-citizens. The Bush administration actively recruited non-citizens by establishing a fast track naturalisation process in July 2002 as part of its war on terror (Mariscal, 2003). Unlike the situation during Vietnam of *alien* conscription where the Australian state compelled through the law the making of soldiers to defend its interests, the war in Iraq is marked by this move from conscription to recruitment. The targeted recruitment of Latinos/Latinas began in the Clinton years when the army was not meeting its enlistment quotas and the then Secretary of the Army realised that Latinos were the fastest growing military-age segment of the US population (Mariscal, 2003). Acting upon such findings the Pentagon launched a massive recruiting campaign targeting the Hispanic Market. A large number of military recruits are signed up through the Delayed Entry program (DEP) which operates in high schools across the nation (Davis, 2007). The well crafted messages on the DEP website have been in development ever since the draft ended and the all-volunteer military was initiated after Vietnam (Davis, 2007).

Jesus Suarez was one such recruit. He was pursued for two years by a US army recruiter, by which time Jesus had talked his parents into selling their home and business in Mexico and moving to California. Jesus enrolled at a school for academic achievement but the recruiter wanted him to transfer to a more troubled school since its requirements were fewer and according to their logic he would graduate sooner (Davis, 2007). Three years later at the age of 20 his body was torn apart in Iraq by an American made fragmentation grenade during the first week of the invasion (Davis, 2007). In this war, citizenship is being used as a recruiting tool aimed specifically at young immigrants who are told that they will be able to quickly get citizenship for themselves and for their entire families even though this is not always true or possible (Davis, 2007). For some, a growing number in fact, citizenship is attained in death, posthumously, making death the only possible way to have been considered after the fact of one's life, worthy of having belonged to a world that one is no longer a part of (Batalova, 2008). Joseph Pugliese, in the context of the US military's recruitment of non-citizen subjects, has commented on the practices of posthumous or "surrogate white citizenship" and from this has developed the concept of "prosthetic white citizenship" (2007a: 15).

> Prosthetic white citizenship is what is conferred upon non-white subjects of the white nation. As a prosthetic, it is a citizenship that cannot be corporeally owned or nativised as the prosthetic of white citizenship remains visibly an adjunct to the non-white body. Understood within the doxic binaries of common sense epistemologies, prosthetic white citizenship is not of the body; rather, as *technè* (technology) in opposition to *physis* (the natural), it can never be corporeally nativised. Prosthetic white citizenship, as a technology of power, is always imposed or conferred from the outside (of the body); as white *technè*, it can only ever be taken up by its non-white subjects as simulation, precisely as a type of prosthetic limb that always gives away its adjunct status as non-native artifice. Even as prosthetic white citizenship can be conferred upon non-white subjects, it can, precisely because it is viewed in terms of an artificial adjunct to the non-white subject, be withheld and erased (Pugliese, 2007a: 15).

In the context of the non-citizen deaths that have taken place in Iraq and have been granted posthumous citizenship, citizenship is finally, not withheld precisely because the non-white body has been literally erased. Pugliese's conceptualisation of the prosthetic invites the question: from where does the power to "confer", to "withhold" or to "erase" citizenship derive? And if prosthetic citizenship can be conferred to those whose status is *alien*, does that then make those who have the power to confer, native? The assumption of the position of native or Indigenous is what propels the power to confer, withhold or erase the right to belong. In this context, the technology of the prosthetic can be understood, at least in the Australian context, as a technology of white sovereignty, one that naturalises and indigenises the violent and illegitimate locations from where such power derives.

In his personal memoirs about the events of the military coup that had taken place on 21 April 1967 in Greece, Andreas Papandreou wrote:

> It must be expected… that this book will anger a good many people. It is written, however, in the hope that, along with anger, something else may be stirred up: an awareness that imperialism, though it was officially buried, is

> not dead in the world. It stalks, as it did before, the struggling nations of the world. Greece has been one of its victims. And unless, through knowledge, something is done, and done quickly, there will be many others (1971: 28).

This essay, which was sparked by a random comment over Christmas lunch, has traversed the northern and southern hemisphere to narrate a transmediterranean history that becomes possible only when imperial and colonial histories of war, poverty and migration are seen as intersecting. The story that has been told is not set purely in the past as the conditions that brought it into being persist. Colonialism and imperialism persist and their language is, more often than not, the language of law. My father's story did not at first appear to me a story about the law. It seemed a story about fate, chance and luck. But at every point in this story I am struck by the violent and covert manner in which law underwrites the fate and luck of people who find themselves poor, war ravaged and colonially controlled.

The Drawing of the Sovereign Line

Lara PALOMBO

Affiliated Fellow at Macquarie University, Sydney

Introduction

Very recently my grandmother had a stroke while staying in a nursing home. Afflicted by dementia and unable to look after herself, Olga had been living in a nursing home in the southern Italian region of Molise. During this time my family, largely living in the northern region of Lombardy had been trying to get her into a publicly funded nursing home in a location closer to them and specifically nearby the city of Como. But the regional regulations that govern Lombardy ruled that because Olga had not been living in this area continually and for a specified number of years she could not apply for a place in a nursing home. Let me be clear here: if she wanted to move to Lombardy she could still move there; so the focus was that this was not to occur under the auspice of the local government, as both Olga, as 'a non resident,' and her children, 'as residents,' were perceived as outside the territorial logic that regulated the provision of services for 'residents' within the region. This provision has elevated a border that draws a regional sovereign line between southern and northern lives. This line follows the logic of a Territorial form of Welfare (officially named as Territorial Welfare) through which, although "constitutional provision for the universalization of health care and social insurance legislation… remain a competence of the state", a great deal of regional autonomy is created for the management and implementation of social policies (see Bagarani, Bonetti and Zampino, 2007; Bull, Newall, 2004; Roux, 2008; Fargion, 2005). In this policy setting, 'national sovereignty' through Italian citizenship is not enough to access services, which are reserved for the residents of the region. But also that 'residents' of the region including Olga's children who embody inter-regional connections with the southern regions are not able to assert these links in moments of need. Ruzza and Scmidtke's explanation of the *Lega Lombarda* explain that through this logic "a Southerner who had lived in the North for a sufficiently long period could become a Lombard provided he or she accepted the values of the new culture (Ruzza and Schmidtke, 1996: 183). Territorial Welfare thus elevated a border that was informed by a regional form of sovereignty or 'autonomy' that prevented altogether Olga, as an Italian citizen from the southern regions, from crossing the border as she was too sick to be moved anywhere else but into a nursing home. In a few words, the border declared where Olga should live and ultimately die.

But if I knew Olga well, this would not have been news to her. The sovereign logic that informs this border has certainly shaped Olga's corporeality and that of her children for a long time. In fact, the mobility that informs this border, or what Rajaram and Grundy-Warr (2007) have summed up as the "borderscape" has long become a norm in the lives of many southern subjects who are consigned to occupy allocated spaces within and outside the northern regions. They have come to embody this border. This is indicated also by the answer that my father gave me when I curiously asked why he and Olga returned to live in the South and separated from the rest of the family in the mid-1970s after migrating to the region in the 1950s; I was told: "because when there is an economic crisis in the North nobody helps you" (Interview, 2009). This is also the same reason that my father used when he explained in 2002 that he was leaving Milan again and going back to live in the South "because there was an economic crisis". This is a lived knowledge of the way the regional border bans southerners from regional sovereignty in Lombardy and locates them in the region under specific northern terms, as a provisional source of labour, always in movement or in transit, ready to leave, back and forth between the South and the North and vice versa or if legally residents, only by not claiming their links to the South. The South thus becomes what Rajaram and Grundy-Warr name as the 'outside space' of exception that lives in a continual relation with the inside north (Rajaram and Grundy-Warr, 1997: xviii). This border relation is a norm.

The anti-southern logic of Territorial Welfare as a borderscape is transformed in its encounters with the *extracommunitari* including refugees and asylum seekers. Soguk argues that borders acquire their meanings contingently, through the knowledge and practices that inform them, thus it is understood that borders are always ephemeral, never eternal: "borders can fold inwards, enveloping and containing individuals and groups in societies within particular regiments of governmentality. Or they can fold outward, restricting entry and expelling irregular migrants" (Soguk, 2007: 284).

The territorial ban imposed through Territorial Welfare complements the National and European Union (EU) technologies of security elevated against the arrival of refugees, asylum seekers, Romani people, and migrants. Thus Territorial Welfare extends its anti-southern logic by complementing the Bossi-Fini law and the more recent practices introduced by Berlusconi, including the so-called 'push-back' policy, which are partly endorsed by the EU. These borders, operating at regional, national and transnational levels become indistinguishable from one another and produce a hegemonic form of whiteness that violently bans the movements of what Pugliese calls the "Souths of South/the South of the North" (Pugliese, 2008: 35), represented largely by migrants and refugees from Turkey and Iraq, Pakistan, Sri Lanka and African countries including Liberia, Eritrea, Somalia, Nigeria (cited in Kosic and Triandafyllidou, 2005: 15). Southern Italians are implicated and take part in the declaration of these sovereign relations.

In this chapter, I want to trace the continuance of certain racialised forms of sovereign grammars that produce a white northern form of Italian sovereignty. I argue that this form of sovereignty is based on the banning of southern desires for self-recognition and that of migrants, refugees and asylum seekers. This chapter considers the ways Territorial Welfare is a form of regional governmentality that reconfigures white northern sovereign relations that have restricted and banned altogether residents with southern allegiances. The banning of citizen's access to welfare across the space of the nation, reconfigures a form of sovereignty that is based on the historical exclusion of southern subjects. This chapter connects the sovereign relations that have historically embodied southern subjects with the current treatment of new arrivals, refugees and asylum seekers. It argues that regional, national and transnational sovereign relations combine to reconfigure this northern form of whiteness but also to produce a hegemonic form of whiteness that implicates southern subjects and Lybians in violent, racialised practices that have historically also excluded them from sovereignty. Finally, the chapter shows how sovereignty was historically re-inscribed in the transnational diasporic context of Australia.

The Southern Question

The fact that Olga will not receive welfare support did not stop her from questioning what had happened to her national sovereignty as an Italian subject. A question that links the present to the past or to the unification of Italy under the premise of producing a universal, modern, citizenship within the nation. This question is also an attempt to intervene in what Edkins and Pin-Fat call the "drawing of the sovereign line" or the refusal to allow the re-formation of the North/South divide at the regional level. But I am not arguing that this "desire" is a "radical and dismantling practice", rather in agreement with Edkins, I also recognize that this is a recourse to the "master signifier" of national sovereignty that constitutes the production of the subject "that evokes it": "once sovereignty is in place, an ethico-political challenge in the name of an alternative becomes illegitimate… Indeed, it functions to define politics in a particular way such that sovereignty is the only referent by which one can understand the political" (Edkins and Pin-Fat, 1999: 7).

In Olga's life, national sovereignty sustains the production of the Territorial Welfare of Lombardy and its embededness in sovereign relations against southern non-residents. The evoking of the national in itself is Olga's reminder of what Pugliese calls "the violence of enforced unification on exclusively northern terms" that historically has always excluded southern forms of sovereignties (Pugliese, 2008: 7). As my father said, "in times of crisis nobody helps you in the North".

Territorial welfare, through its regional borders, as Rajaram and Grundy-Warr put it "continues to clarify the boundaries of the norm" (Rajaram and Grundy-Warr, 2007: xix). This norm is a configuration of the nation in northern terms that excludes southerners from sovereignty. This exclusion re-produces the effects of the colonial occupation of Southern regions during the unification of Italy. This was headed by a white sovereign form of power stemming from the northern Piedmontese Kingdom that, as Ressa shows, was eager to strengthen its own sovereign powers and take a place among the European nations as a trading partner in the Mediterranean. It is known that this kingdom was also "on the verge of bankruptcy" and that the conquest of new territories was a way out (Ressa, 2008: 50-55). The unification of Italy, according to Gramsci's primary text, *The Southern Question*, guaranteed the unification of capital across Italy but also the perseverance of feudal and restrictive economic conditions that excluded land redistribution (Gramsci, 1975; see also Pagano, 2002). The formation of the nation-state was therefore part of a white northern colonial practice that excluded selected sovereign demands from the south to maintain its exclusive possession of the newly formed nation-state. This process was sustained through a violent military occupation of the South that created a ten-year civil war, which saw one million people being murdered by the Italian Army (Pagano, 2002). The zone of the South then came to operate as a colonial 'zone of exception' where forceful northern military intervention against the so-called "brigandage", that as Salerno's recounts is "the term by which the armed Northerners attempted to defame the [South's] rebellious peasants against the invasion" (quoted in Pugliese, 2008: 1). The sovereign power of the newly formed nation-state under northern rule suspended the law and created a "terror formation" within the southern zone. This, as Pugliese has argued, also rendered the south coextensive with colonial discourses on Africa: the colonial concepts of the "terra incognita" and "dark continent" begin to resonate with a northern European model of colonization. In this setting,

"Africa in fact becomes the governing metaphor through which Northerners made sense of the South" (Pugliese, 2008: 3).

From the moment Olga was born, she knew she was identified as a Southerner. It is appropriate here to apply Moreton-Robinson's critique of a western form of whiteness that as a regime of power exercises a political hegemony or sovereignty. The possessive logic of a white northern form of sovereignty operated here as an "ontological and epistemological a priori" that framed the configuration of the Italian nation (Moreton-Robinson, 2004: 75). This form of whiteness operated to "naturalise" at the "level of beliefs", and discursively at the level of epistemology, the nation as a white northern possession. The possessive logic of white sovereignty "naturalised" the position of women from southern regions. This process of naturalisation concealed the violence that produced national sovereign authority through a politics of subjection of the subject that carefully elevated white northern sovereignty (Edkins and Pin-Fat, 1999). Petraccone's analysis of the correspondence sent by the northern soldiers and officials located in the south during the unification recounts how women from Molise were embodied as: "backwards murderous women" that were said to "tie gentlemen from the testicles and... pull them like that in the streets..." (quoted in Petraccone, 2000: 28). This representation in effect produced a "public opinion favourable to the Piedmontese" who were differentiated from the "isolated", "barbaric" conditions and practices believed to engulf the southern regions and their people (Petraccone, 2000: 28). In this setting, the northerners through unification and national citizenship were represented as improving women's status. The benefits of nation building and national unity were even preached by educated British middle class women (at times feminists) who travelled to southern Italy, especially after the 1860s, "as women who sought to assist in the development of the Italian nation, and the turning of its people particularly women, into citizens" (O'Conner, 2003: 253).

In the late 19th century, the work of the northern scientist Cesare Lombroso, naturalised further the superior position of Northerners by embodying them as the norm against which the southern subjects were measured, thus re-affirming the sovereignty of a northern form of whiteness. This whiteness operated as an ontological and epistemological a priori that produced what Edward Said calls an "affiliative order" that represented the "hegemony of the dominant culture" (Said, 1983: 19-23). Lombroso's work constructed "new forms of cultural relationships" or "affiliations" that were:

> produced in complex ways by modernism, filiations gives birth to affiliation. Affiliation becomes a form of representing the filiative processes to be found in nature, although affiliation takes validated non biological social

> and cultural forms… This keeps relationships within the narrow circle of what is natural, appropriate… and thereafter excludes… all the political dimensions (Said, 1983: 23, 23).

Lombroso, through a series of measurements and taxonomies of women's bodies, affiliates their crimes to their regions of origins, especially that of women "from the [southern] country districts and… [from] the Islands" of Italy (Lombroso, 1911: 28). These were based on a comparative analysis of the data that located these regions within the North and South divide and differentiated them according to this division:

> It is apparent, then, that these crimes are most frequent in the provinces where the population is predominantly Semitic (Sicily, Sardinia, Calabria) or purely Latin (Latium Abruzzo) as compared with those where the population is Teutonic, Ligurian, Celtic (Lombardy, Liguria, Piedmont), or Slavic (Venetia)… It is, then, to the African and oriental elements (the Greeks excepted) that Italy owes the frequency and homicide in Calabria, Sicily and Sardinia; while the occurrence of a small number, as in Lombardy, is due to the large Teutonic element in the population (Lombroso, 1911: 28).

This is what Said calls a form of "horizontal affiliation" that produces "women with… new forms of relationships" that not only connect women's crimes to their perceived biology but also to that of other racialised and sexually differentiated women's bodies, embodied here as "Negro woman", "Indian women", "Hottentont African women", "Abyssinian Women" and "Australian Aboriginal Women" (Said, 1983: 19; Lombroso, 1895: 113). The racialised sexual characterisation of the "primordial" southern female subjects is then embodied through forms of affiliations with other women from the Souths of the South. In this setting, the northern women who are referred to as "modern" are rendered invisible or assumed to be the 'norm' when contrasted to the characterisation of the "primordial" women, but are also clearly marked as being outside these racialised and sexually differentiated affiliations; their whiteness remains unmarked. Afflicted by the effects of this continual violence, Olga in the late 1950s decided to migrate to the North to improve her economic conditions. Bio-political technologies regulated Olga's provision of both her labour and that of her children to the region of Lombardy that, after the war, "had an imbalance between low rate of births and high rates of death" and most importantly needed a new labour force.

Territorial Welfare

It has long been argued however, that the creation of the nation-state has not managed to fulfil northern sovereign interests. Territorial Welfare was produced in the 1990s by northern based biopolitical relations that effaced further the modern promise of national sovereignty for southern lives by demanding a "decentralization" of state power. In the 1990s, as Anna Bull's summary specifically indicates, the *Lega Nord* argued that:

> the 'Southernization' of the Italian state, as evidenced by the emergence of a widespread corrupt and clientelistic system of government [and this] was a clear sign that the process of national unification, far from promoting the homogenization of the country around northern laws, practices, institutions and economy, had succeeded only in imposing 'southern' deviant practices upon the whole country (Bull cited in Fargion, 2005: 140).

Territorial Welfare then becomes a form of territorial governmentality that attempts to reinstate and/or strengthen a northern form of white sovereignty. This is linked to the re-alignment of national party politics after the Tangentopoli scandal and that created a vacuum for the intervention of the *Lega Nord* (Northern League) and for its demand for decentralization, regional autonomy and overall for a new form of federal politics (Fargion, 2005: 140; Roux, 2008). The *Lega Nord* represented several parties of northern and central Italy, and was in favour of a form of regional governmentality that would re-territorialize northern ethno regional sovereignties within the nation-state: "our chosen form of autonomy had to be ethno federalist i.e., a union of several ethno national entities in a unified political instrument capable of victory. Not isolationism but a fight against the centralized power of the state" (quoted in Clark, 2008: 392).

This call for regional sovereignty included declaring restrictions to the allocation of local services to southern workers but also immigrants, refugees and asylum seekers (see Ruzza and Schmidt, 1996; Tambini, 2001). Thus this focus on a northern form of (white) sovereignty has reiterated historical racialised accusations against the South being backward, expensive, clientelistic and corrupt and one that benefited from northern richer resources (Roux, 2008). And more recently by attempts to ban service provision for refugees and asylum seekers as for example shown by the *Lega Nord* decision in October 2009 to lobby at the regional level to ensure that only 20% of *case popolari* or public housing will be distributed to *extracommunitari* (*Lega Nord*, www.leganordromagna.org/attualita-cultura/case-popolari.php). In this case, through the re-territorialization of the function of the state and the region the modern category of 'citizenship' has been reconfigured in ways that cannot be applied across the territory of the nation.

The EU has also supported the production of territorial forms of governmentality. Part of the Italian state's support for this form of 'regionalization' is linked to the formation of the European Union (EU) and its commitment to neoliberal practices and the re-drawing of EU internal and external borders. The Italian participation in the formation of the EU has in fact been formalised through the state participation in processes of 'Europeanisation' of territories that aim to create social cohesion amongst diverse nation-states. Financial initiatives such as for example the *Fondo Sociale Europeo* and the *Equal: Iniziative Comunitaria*, directed at "the Europeanization" of territories" has established the re-territorialisation of the national subject by emphasising their local values in producing socio-economic benefits (www.fondosocialeeuropeo.it). As it has been explained the Equal initiative is: "a real valorization of the territorial subjects… so to give life to virtuous networks capable of producing resources, build alliances between the public and the private, profit and non-profit and to manufacture responsive answers to the needs of weaker sections of the population that reside in the territory" (EQUAL, http:// http://ec.europa.eu/employment_social/equal/index_en.cfm).

This re-territorialization for Fargion has helped the states to address the "emerging and costly needs related to industrial restructuring, diminishing rates of return in the economy and… fiscal austerity" by decentralizing a number of fields in social policy: means tested public assistance, social care services for children and the elders, labour market policies and health care (Fargion, 2007: 3). The EU then has altered the centrality of the modern state in the management of provisions of social protection schemes (i.e., national provisions for social security benefits). But in this process, as Ferrera argues, the EU has also redrawn the "boundary of welfare" per se:

> European integration can be understood as a large-scale operation of boundary redrawing: the redefinition or removal of state national-boundaries within the EU space in respect to an increasing number of functional spheres and institutionalized practices, including social sharing. In the wake of the EU's free movement and competition rules, the nation state is not longer the sole and ultimate arbiter of inclusion and exclusion into its own redistributive spaces… Given the importance of social sharing for material life chance, cultural identities, and legitimization dynamics, reshuffling the 'boundaries of welfare' may destabilize the basic architecture of Europe's national societies and political systems (Ferrera, 2005: 3).

So welfare through decentralization becomes linked to an EU process of "redrawing internal and external boundaries" associated with the European unification and the state and by so doing has extended the function of Territorial Welfare at a regional level.

Drawing on Fargion interrogations of the way territorial practices may undermine social cohesion (Fargion, 2007: 5), I ask here what happens when territorial competition for welfare services intensifies with the arrivals of migrants, asylum seekers and refugees? And what happens to the provision of welfare if the territory is based on the sovereign logic of a northern form of whiteness predicated on the production of the North/South divide and the banning of the Souths of the South from regional sovereignty? And again, what happens to welfare when the state and the EU intensify their border protection policies to ban the movement of new arrivals? Ferrera, in 2005, already warned that this "regional sovereignty" has already created in countries like Italy an "inward" regionalism: "[this] tends to be accompanied by explicit proposals/demands to gain control over the territorial entries of foreign immigrants and their access to welfare benefits and services" (Ferrera, 2005: 235).

The territorialisation of welfare ultimately produces internal borders that reconfigure the racialised anti-southern logic of white northern sovereign relations and extends it against the Souths of the South of the North.

The Refusal to Draw the Sovereign Line

The logic of Territorial Welfare enacts regional, national and EU sovereign relations that demarcate access to basic living resources such as food, health, housing and education at the local level. This is demonstrated by the response received by a group of 300 refugees and asylum seekers during a recent occupation of an empty building in the city of Milan. The 300 legally recognised refugees from Eritrea, Ethiopia, Somalia and Sudan, consisting of families and individuals, through this occupation attempted to bring attention to their poor and unsafe living conditions and overall to the limited regional, national and EU support for their status. The focus centred on the lack of permanent forms of accommodation and public housing for refugees. These subjects had been living for months on the streets, with limited if any form of assistance (Rifugiati di Milan, 2009). Thus the occupation became a protest against the treatment of refugees and asylum seekers and an attempt to refuse the ongoing effects of the violence that had also brought them to Italy. Sovereign power for Edkins and Pin-Fat is a form of violence (sovereign violence) that can be challenged in two ways: "first as a refusal to draw lines between forms of life; and secondly, an assumption of bare life… In the end, the refusal to draw lines and the assumption of bare life seek to reinstate properly political power relations with their accompanying freedoms and potentialities" (Edkins, and Pin-Fat, 2005: 1).

In this case, this occupation operated as a refusal to draw lines between forms of life. This was an attempt to intervene in "the drawing of the sovereign line" that repeatedly positioned these lives outside national and regional forms of support available to citizens and residents and that by so doing has legitimized and continued the abuses confronted by these subjects (Edkins and Pin-Fat, 2005). During the occupation it was stated: "We are refugees… we are in a difficult situation, we are asking that our rights be respected. Today they have beaten us like animals. They do not want to do anything for us; we will fight until the end of our lives" (cited in Nicola Grigion, Progetto Melting Pot, 23/4/09).

But with these refugees the potentiality of refusing the drawing of the sovereign line was undermined by the formation of a hegemonic sovereignty operating at the national, European Union and regional level which is also supported by many Southerners. In this case, the protesters demanded the intervention of the EU, which after meeting with the protesters argued that it could not help them and that it had already provided adequate funding to the Italian Government. The Council of Milano, on the other hand, agreed to provide beds for 8 months in dormitories under the conditions that:

- women and men would be separated from one another, as there were no dormitories for families,
- during the day (from 8 am to 6 pm) the occupants would leave altogether their accommodation.
- This was not considered an ongoing form of accommodation

(Rifugiati Di Milan, 2009/04, rifugiatimilano.blogspot.com/2009/04/cronistoriasempreaggiornatadelle.htl).

These are the same rules that prevail in the current five *Centri di Accoglienza* (reception centres) located in the city and that also led to the 300 refugees being stranded without accommodation. The point was that these dormitories represented a refusal to provide permanent forms of accommodation. This logic was also supported by the Berlusconi Government's laws that, since 2008, have legislated to prevent *extracommunitari* including recognised refugees, from applying for public housing if they have not been in a region for five years or within the national territory for 10 years (Rapporto Congiunto Open Society, 2009). Within 15 days, the 300 refugees were again on the streets without any form of accommodation. This event then re-asserted at the regional level a hegemonic northern form of whiteness that combined regional, national and EU borders to ban the 300 refugees from accessing permanent forms of accommodation. This not only ignored the gross violations faced specifically by refugees, but it also overall reduced these subjects to 'bare lives' banned from the security of decent living conditions.

White Proximity

Residents of the southern regions are evoked as sovereign subjects by regional, national and transnational immigration policies and practices regulating new arrivals within Italy and the EU. So whilst as I have argued above that Territorial Welfare in the region of Lombardy effaces Southerners' national citizenship and sovereignty, the latter is also reconfigured through national, transnational and regional borders operating within the southern locations of Italy in order to produce the ban of "illegal" migrants, *clandestini* and *extracommunitari*. These borders are enacted in the South of Italy to produce what Agamben calls the "modern biopolitics of the Sovereign subjects" who decide the value and non-value of life (Agamben, 1998: 142). It is clear that through this enactment many southern subjects become direct participants in the banning of new arrivals from sovereignty. Pugliese also remarks that many are directly benefiting from the exploitation of the *extracommunitari* as the "south is marked by its own internal history of racism" (Pugliese, 2008: 25; Pugliese, 2007). These internal histories of racism become coextensive with and indistinguishable from the current regional, national and transnational demand to adopt forms of territorial sovereignty to ban new arrivals and re-produce a hegemonic form of white sovereignty. The recent, statistical data from 2007 produced by the *Ministero del Interno* (Minister of the Interior) indicates that the region Molise, in 2007, only offered two residential structures for *extracommunitari*. And that Molise was not alone in this as other southern regions also offered very limited services. Also, as recounted in a recent Report by Amnesty International, the limited availability of regional services in zones of the south are accompanied at times by violent attacks against new arrivals:

> On 13 May, up to 100 people, reportedly armed with sticks and Molotov cocktails, set fire to parts of a Romani settlement in the suburb of Ponticelli in Naples. One Molotov cocktail was thrown at a trailer housing a number of children, who only just managed to escape being burnt alive. In total, around 800 Romani people were forced to flee the settlement. On the same day, several Romani people were also physically attacked in the surrounding area (Amnesty International Report, 2009).

These practices occur within the grammar of sovereign relations that implicates (Edkins and Pin-Fat, 1999) southern subjects in a violent politics that is not only based on "operation of law but also on hierarchies of bodies" (Rajaram and Grundy Warr, 2007). These practices in effect produce a "racialised recalibration and re-assignation of Southerners" as citizens of Italy, the EU and residents of southern regions (Pugliese 2008: 24). In the context of unequivocally non-white "outsiders", southern Italian subjects that historically had been identified as non-white now get elevated "up the white supremacist racial hierarchy" toward what Pugliese calls a "proximate whiteness". Then, if we follow this logic, the co-extension of national, transnational and regional borders within the southern regions (i.e., Territorial Welfare), brings Southerners "politically and bureaucratically within the proximate folds of whiteness" (Pugliese, 2008: 25). But this "proximation" as Olga's bare life indicates does not undo the drawing of the north/south divide and the consequential "precariousness" associated with many southern lives. The borderscape produced through hegemonic whiteness includes and implicates Southerners in a violence against the *extracommunitari*.

The Treaty on Friendship, Partnership and Cooperation

The relation between Libya and Italy is one that I want to discuss further here, in the context of ongoing sovereign relations between the countries that were marked by Italian colonialism. Following the signature in August 2008 of the Treaty on Friendship, Partnership and Cooperation between Italy and Libya, Berlusconi claimed that this Treaty puts an end to the long-lasting colonial dispute by granting both material compensation to Libya and launching a strategic economic and anti-immigration partnership. Libya, between 1911 and 1943, was subjected to what Mbembe calls the "terror formation" of the colony where white Italian sovereign forms of necropolitics produced a systematic violence that included killings, wounding, sexual violence, starvation, forced exiles, dislocation, dispossession, environmental destruction and more (on necropolitics see Mbembe, 2003; on violence in Libya see Salerno, 2005; Del Boca, 2003; Piper, 2002; Gazzini, 2009). Salerno's book *Genocidio in Libia* notes the ways the Italian government "was not particularly worried with the deaths of Libyans", and the testing of new military technologies including bombs and poisons were ways of "killing and terrorising Libyans" but also testing grounds for other colonial expeditions (Salerno, 2005: 119). One of the many devastating factors of this colonial invasion is the way white Italian sovereign relations, in the war against the Resistance movement that re-organised itself in the early 1920s, came to dislocate entire populations of civilians and concentrate them in new areas such as Syrtica and Cirenaica that eventually were erected as concentration camps (Salerno, 2005: 87-88). As it was recounted by the Italian Army Health Department, Dr. Todesky, "By the end of 1930 all Libyans who lived in tents were forced to go and live in the camps. 55% of the Libyans died in the camps" (quoted in Piper, 2002).

In the 2008 Treaty, Berlusconi declared that "a painful chapter of history is finally over" (quoted in Marchetti, 2009). The end of this "chapter" is linked to the injection of "funding – up to $5 billion over the next 20 years – for key infrastructure projects, the nature of which is left undefined. Libya will have the right to propose what type of infrastructure projects it is interested in" (Gazzini, 2009). Although the Treaty produces economic benefits for Libya, the claim to wanting to end past disputes, whilst simultaneously implicating Libya in the establishment of violent technologies of security, whose prototypes were also tested and used against this colony to be used against "illegal" migrants raises many questions. The following translation is a quote from an interview with Seif el Islam al Qaddafi, the son of the Libyan leader that explains that: "the relations between our two countries are very deep and important… but in the relations between our countries exist two unresolved problems: for us is the question of reparation and for Italy illegal immigration" (Salerno, 2005: 130).

And I want to reflect on the way the Italian Government, with the help of the EU, has engaged with the "ending" of this unresolved problem. I see a connection here between the production of Libya as a border of security and the historical attempt by the Italian Government to deal with the failure of Italian unification to produce better living conditions including land distribution in the late 19th century. In this case, colonialisation of countries like Libya was envisaged partly as a solution to guarantee landless classes of southern Italians with land, without interfering with existing northern-southern alliances. In 1893-1894, the Sicilian born Minister and Premier Crispi responded to the *Fasci Siciliani* that were demanding land redistribution: "the assignment of the current generation was to give the country a consciousness of itself" and "to make Italy equal among equals" or an equal amongst other European countries, which meant, in his estimation, that that was the most difficult task since unification (Cunsolo, 1991: 91). Modern "equality", under Crispi, took the meaning of developing a national capacity to trade with other European countries by embracing exploitative forms of economic nationalism that were to include the project of colonialisation within the South but also abroad. And the South was to become a source for the provision of the necessary labour force to colonize Somalia (1891), Abyssinia (1887-1941), Libya (1911-1932) and Ethiopia (1935-1936) (see Pugliese, 2008: 8). The benefits associated with these violent processes, however, did not undermine the historical production of the North/South divide. I argue that the current relations between Libya and Italy reconfigures in this context a similar historical logic that claimed to respond to the effects of the northern occupation of the south by implicating southern Italians in the project of colonisation. This treaty is a remapping of this colonial sovereign logic that implicates Libya in the production of a sovereign violence against southern refugees (against the Souths of the South). The Treaty responds to Italian violent colonial history by implicating Libyans in its violent sovereign ban of refugees and asylum seekers. Libya through this remapping is then transformed into a border zone of exception that is both outside but also inside the jurisdiction of Italian sovereignty.

The Treaty clearly sets up a new regime of biotechnologies of security against illegal migration. The Treaty's new detention centres, deportations to and from Libya, and joint Italian-Libyan police patrolling of the Libyan coastline become a part of the white hegemonic sovereign relations that also include the European Union, which is financing some of these technologies (Ronizetti, 2009; Andrijasevic, 2005). It is relevant here to consider how Suvendrini Perera discusses the so-called Pacific Solution produced in Australia as a "remapping of new topographies" of historical hegemonic sovereign relations that are based on:

> inclusion and exclusion on land and sea by this exercise of sovereignty over refugee bodies: the contraction and expansion of national borders as detention camps are placed outside the space of the nation and lines are redrawn in the sea to deterritorialize and excise parts of Australia from migration zones while yet other parts are annexed under measures such as the Pacific Solution, and new cartographies of surveillance constituted through extensive militarization and policing of the oceans. All these movements contribute to the making of a racialised borderscape that is continually mapped and remapped through a number of operations (Perera, 2007: 121).

In a similar vein to the Pacific Solution, this Treaty remaps white hegemonic sovereign relations in ways that expand Italian and the EU sovereign relations outside their territorial jurisdiction. Libya operates as a racialised borderscape that, through the elevation of the detention camps and patrolling of the seas, works to ban new arrivals from the sovereignty of the Italian nation and of the European Union. This effectively targets and disconnects certain regions and its people from the Souths of the South from accessing Italian and European resources. Through this process, the Italian state and European Union exempt themselves from their responsibilities to assess and provide assistance to asylum seekers and other new arrivals. Ultimately, these may produce economic benefits for Libya but do not undermine the sovereign logic that led to its colonisation.

Australia and White Diasporic Relations

Returning to Olga's life, I want to ask what happens to a migrant like myself when news of relatives located overseas reaches us? Specifically, what can one do or what could I have done to help Olga? How do diasporic lives come to embody transnationally these white sovereign hegemonic relations? Many diasporic subjects and communities have relatives caught in a war, that are being killed, tortured or lost at sea or inland while escaping war, or that embody the knowledge of poverty and hardship or incoming disasters or starvation or economic exploitation that affect their own communities overseas and relatives. These diasporic lives, although they do not share the same cultural relations to the everyday, they are nonetheless interlinked by the white (diasporic) sovereign relations that rule the distinction between permissible and non-permissible transnational relations. Through the continual recourse to bio-political technologies of National Security and of Population Control, sovereign re-alignments are enacted nationally to regulate the everyday of Indigenous and diasporic lives in ways that identify and distinguish the privileged white diasporic sovereign lives that are somehow dispensed or immunized from the scrutiny that the rest of the community receives in Australia (on immunity see Esposito, 2003). The historical demand for national loyalty to a white (i.e., colonial and anglophilic) diasporic form of sovereignty has been consolidated through racialised regime of bio-political technologies that have attempted to continually regulate the assimilation of new arrivals and Indigenous Australians and more specifically to restrict diasporic and Indigenous affiliations that questioned white diasporic relations.

But the drawing of the sovereign line has also always being questioned by both Indigenous and diasporic relations. In the following discussion, I want to focus on the way diasporic political relations have historically been mediated by Italian sovereign relations when responding to the "drawing of the sovereign" line in Australia. This section focuses specifically on an historical example of the way Italian national sovereignty was reconfigured transnationally through the *Italo-Australian Journal* and how it mediated the newspaper's response to the drawing of the white sovereign line in Australia that overshadowed the banning of southern Italians and southern European lives. This example is in no way perceived to represent the wide spectrum of political responses enacted during this historical moment. This discussion draws on a larger body of work examined in my doctoral dissertation that focuses on the internment of women of Italian origins in Australia during the Second World War, who by becoming perceived as political lives embodying Italian sovereign relations, were posed as a "national threat" to white sovereign relations in Australia. Through their detention in the space of exception of the internment camp they were banned from national sovereignty. The newspaper discussed here was often used by the military authority as form of evidence to intern these women.

The *Italo-Australian Journal*, as a form of representation of diasporic European relations operating in Australia, was concerned with guaranteeing protection for subjects identified as Italians. This was established in Sydney in 1922 as a weekly bi-lingual newspaper available in Victoria, Queensland and South Australia and, in accordance with the law of the time, it received formal approval for its existence. The founder of the newspaper was a merchant, Francesco Lubrano, who worked over the years with a number of directors – Antonio Folli, followed by A.P. Rimoldi, Porfirio Scotto, Eustacchio Del Pin, Franco Battistessa, Fortunato la Rosa (Briani, 1977). My analysis, however, focuses on the newspaper's evocation of a sovereign (transnational) power that becomes productive of the "white possessive individual". The white sovereign subjects deployed in the newspaper are the effect of the diasporic longing for sovereign politics. In this newspaper, the recourse to a transnational form of sovereignty produces the concealment of the violence associated with the national unification and the formation of a national sovereign power in Italy. Its primary role is one that, by sustaining an identification of the people with the nation, comes to naturalise sovereign relations (Vincent, 2002: 34). The *Italo-Australian Journal*, from its inception, was clearly marked as an ethno-nationalist and masculinist diasporic project that became involved in what Ray Chow calls a "strategic political re-affirmation" of a homogenous Italian diasporic identification (see Chow, 1990; Ang, 2001). In its first edition it stated:

> From this first edition we take the opportunity to send a brotherly welcome to all the co-nationals that are living in the lands of Australia; what brings us together is our Nation and this communion facilitates our task, that is chiefly one of harmony, of brotherhood (*Italo-Australia Journal*, 1922: 1. My translation).

This ethno-nationalist and masculinist re-imagining of all Italian migrants as strongly attached to the "brotherhood of the Italian nation" evokes here the capacity of the nation to 'unify' the scattered diasporic subjects in Australia. In this sense, then, the newspaper from its very first edition produces a form of representation that concealed the sovereign violence associated with the forced unification of Italy. As a result, it creates a white diasporic sovereign space that claimed to produce "unity" amongst Italian migrants or a sense of harmony between migrants and the Italian nation. This sense of harmony naturalizes sovereignty as the "natural" condition of the state (Vincent, 2002: 34).

The newspaper re-imagined all Italian migrant subjects as equal sovereign members of the (white) brotherhood of the Italian Nation. The evocation of a "national brotherhood" is signified to produce a form of sovereign protection and support for "all Italian migrants" living in Australia. The transnational diasporic reconfiguration of Italian sovereignty is thus linked to the "protection" of a unified diasporic community abroad:

> We believe today in the necessity of an Italian voice that would defend all Italians living in Australia when required. And it is today more than ever that it is necessary to protect our interests as today the Italian immigration in Australia has taken noticeable and important proportions and is continuously growing (*Italo-Australian Journal*, 1922: 1. My translation).

The phrases here of "we believe and "our interests" attempt to produce an overall sense of "equality" or "unity" within the brotherhood that renders invisible the North/South divide attached to white Italian sovereignty. Thus the evocation here of Italian nationalism reconfigures the knowledge of how these immigrants were constituted through the very failings of national sovereignty. This nationalism evokes, in Mbembe's words, "the romance of sovereignty": "in this case, rests on the belief that the subject is the master and the controlling author of his or her own meaning. Sovereignty is therefore defined as a twofold process of self-institution and self-limitation (fixing one's own limits for oneself)" (Mbembe, 2003: 13).

The newspaper itself is thus configured as a zone where Italian (diasporic) sovereignty prevails. It romanticises the 'harmony' and 'unity' established through the nation by concealing the North/South divide and its consequences. In this form of representation, migrants are disconnected here from the failure of Italian unification to produce better living conditions for them. The growing migration of Italian nationals, as Gabaccia reminds us, indeed followed the unification period: "Over one million applied to migrate between 1876 and 1885; over two million in the following decade; and over four million in 1896-1905. Almost six million applied for passports in the ten years before Italy entered First World War" (Gabaccia, 2000: 58).

The newspaper rendered invisible the fact that, after unification, the emigration from southern regions surpassed the movements from the northern regions (except from Veneto, which had been affected by the limited re-distribution of land and the displacement created by growth of new industries) (Gabaccia, 2000: 69). Olga's father was one of the many who also migrated in the early 1900s to the US.

Sovereign relations operating through the newspaper produced a national form of representation of diasporic women's lives that by rendering invisible the southern question and the north-south divide, effectively denied and concealed the violence that was shaping their lives in Australia. This becomes clearer in the *Italo-Australian Journal* appraisal of the final findings of the Ferry Report (1925) which was part of a national inquiry into the arrivals of southern Europeans and that included questions over the desirability and whiteness of Italians after that various disputes erupted over their working practices. The Report argued that Northern Italians "were absorbed in the social and economic life of the country" (Ferry quoted in Douglass, 1995: 141). They were embodied as "settling in well, establishing families and that their children were becoming good Australians" (Douglass, 141: 1995). In contrast to this, the Sicilians like the Maltese and Greek migrants are described by Ferry as "illiterate, impoverished, and... less likely to be assimilated" (Douglass, 1995: 141). Commissioner Ferry overall was "less favourably impressed with the southern Italians, who constitute a majority of the new arrivals during the latter years... The South Italians are more inclined to form knots and less likely to be assimilated" (Lyng, 1935: 104). Here the racialisation of the "South Italians" as displaying a diasporic desire to establish collectives or communities that shared cultural practices are conceived to be signs of a failure to assimilate into dominant whiteness. This establishes an anxiety over the whiteness of southern migrants and that of their children. In its response to these findings the newspaper wrote that:

> All Italians... will welcome the Report from Mr T. A Ferry... We do not dispute that from Australia's point of view Italians are aliens. But there are aliens and aliens. It is not right for a cultured, clever, industrious, thrifty and scientific race like Italians to be grouped together with all the other races other than British as "alien", when the word is used in conjunction with migration to Australia. Without wishing to wound the feelings of such people as the Greeks, Maltese and others, we declare (and the facts support us) that Italians are a superior people and are suitable for settlement in Australia. Italians settling in Australia become Australians. Mr Ferry's report is valuable if only because it distinguishes between the different races. (*Italo-Australian Journal*, June 13, 1925)

The newspaper, in effect, through its production of a unified notion of a superior Italian race, reproduces the logic that informed the Ferry Report. They both re-configured notions of a white diasporic sovereignty that are based on a northern form of whiteness and that drew the sovereign line between lives. According to this shared logic then, both the Report and the newspaper assert the racial superiority of Italians by referring to a racialised "proximity" to the whiteness of northern Europeans. This disconnects Italians from southern Europeans but also from the south of Italy. The newspaper, by supporting the banning of Maltese and Greek migrants categorised as southern Europeans, conceals the violence exercised specifically against southern Italian migrants operating within the Ferry Report itself. This strategically responds to the Report by locating the "Italian race" within the proximity of a northern whiteness that, however, reasserts the originary exclusion of southern Italian lives from sovereignty. The newspaper's investment in northern whiteness ultimately produces a silence within the newspaper on the lives of southern Italians within Australia. This creates a diasporic zone that conceals the violence of the white Australian policy.

The a priori knowledge of whiteness that informed the newspaper and the Ferry Report in effect both normalised and extended the current restrictions, exclusion and forced deportation of subjects identified as non-white, non-Europeans and as Aboriginal people. These cultural productions participated in what Perera calls the "anxious hierarchies around intricate calibrations of race, work, and levels of assimilability" that "are worked and reworked in exchange with the specific fears and desires of white supremacy and Australian nationalism (Perera 2008: 279). The desire for proximity to white supremacy within Australia predicated within the Report and the newspaper was based on enforcing the banning of Indigenous and non-European sovereignties and came to be extended to include selected southern Europeans defined as "non assimilable", including southern Italians. The sovereign relations exercised within the Italian newspaper by supporting the Ferry Report established a hegemonic transnational form of whiteness that attempted to extinguish Indigenous sovereignties and to maintain the Australian nation as a northern white possession (see Moreton-Robinson, 2004).

This essay has conducted a personal journey in various locations that have been marked by a northern form of whiteness. It started as a personal reflection on how the modern promise of national sovereignty through citizenship has never guaranteed access to sovereign resources to my grandmother. Olga's relationship to the omnipotent *Nord*/North has repeatedly affected her status as a citizen and shaped her life and that of her children. This is a relationship that has been based on a historical racialisation of southern subjects during and after the enforced unification of Italy that violently asserted the sovereignty of a northern form of whiteness. This is also the same relationship that has shaped Olga's last few days of her life, when the Territorial Welfare of Lombardy decreed that her citizenship was not enough to access the region's health resources and by doing this effectively declared where she should die. The essay then proceeded to a critical analysis of how processes of territorialisation elevate regional, national and transnational borders that become distinct but also indistinguishable zones of sovereignty that come to operate within the same space simultaneously. I have argued that the mobility of these territorial borders or the borderscape produce a hegemonic form of northern whiteness that is informed and based on all these sovereign relations in order to forcefully divide and ban lives. This essay has marked specifically how sovereign relations have repeatedly, and at different historical moments, racially marked southern Italians living in the North of Italy, diasporic subjects in Australia and, more recently, *extracommunitari*, including migrants, refugees, asylum seekers and Roma people. Violent sovereign relations operate to both ban and restrict the movement of targeted racialised groups, especially as they refuse the drawing of the sovereign line. This process locates lives in prescribed places, outside of sovereignty. And when the ban is lifted, these borders can still intervene at local levels by regulating these lives' access to essential local resources including health services and housing. Finally, regional, national and transnational sovereign relations have implicated historically racialised others, namely southern Italians and Libyans, to enact violent sovereign intentionalities against new arrivals from the Souths of the South of Italy. They have now been asked to participate in the elevation of a hegemonic form of whiteness and to embody the border zone. These practices, however, do not undo the white northern relations that inform its sovereignty.

Giving the Voiceless a Voice

Saving Third World Women Through a Western Lens

Seren DALKIRAN

Postgraduate Research Student,
University of Sydney

Introduction

East/West – this division is far too easily drawn, particularly in contemporary debates. Political developments of the last few years, such as the so-called 'War on Terror' and the 'Axis of Evil' in the Middle-East, the endless integration debates and multiculturalism discourses, as well as the way in which Islam and immigrant minorities in Western societies are often perceived as 'backward and inferior' are all symptoms that confirm and maintain the constructed binaries. All this raises the question as to whether it is possible that contemporary perceptions of the 'Orient,' including images of Islam and Muslims, still stand very much in the tradition of Orientalism. It is remarkable to see that the debates mentioned above are commonly shaped around the position of Muslim women, since it is often perceived that they are subordinate to men. From a Western point of view, the basic assumption is that this does not apply to Western women, who are considered to be already emancipated. However, the question remains as to whether this is a reality; the claim that Western feminism has reached its objective is disputable. In this essay, the perception of women in the 'Orient' as passive and subordinate will be discussed in the context of postcolonial feminism, the practise of female genital mutilation, the issue of the veil and the discourse of the former member of the Dutch parliament Ayaan Hirsi Ali within the framework of honour-related violence. Emphasis will be placed on the paradoxical question whether so-called 'third world' women need to be 'saved' by the 'West'; this issue will be approached from the angle of various practices.

The Third Versus First World Binary

The stereotypical portrayal of 'typical' so-called 'third world women' as helpless victims who must be 'saved' by Western feminists has been challenged by both post-modern and postcolonial studies. Using this approach it should be realized that it is not possible to gain new insights without recognizing the wide range of personal perspectives that 'third world women' experience. The dilemma for Western feminists however still pivots on whether non-Western women (in particular, Middle-Eastern and African women) must be 'saved' by the West; and, furthermore, whether gender equality and liberation of women belonging to these specific groups can be achieved whilst still preserving their own culture with its traditions and customs. These crucial questions have an enormous underlying complexity.

In development theory, representations of women have often been a topic of interest in postcolonial and post-modern feminist criticism. For instance, the feminist theorist Chandra Mohanty (1991) believes that large numbers of researchers, especially Western feminist academics, have been inclined to produce one-dimensional, generalizing and essentializing constructions of third world women. Therefore, much of the modern discourse on women and their development "discursively colonize[s] the material and historical heterogeneities of the lives of women in the third world, which produces an image of an 'average third world woman' who is an object of development" (Mohanty, 1991: 53). However, perceiving these women as belonging to the same group is troublesome since this would ignore the differences between them pertaining to cultural identity, religious background, ethnicity and gender identity (Wood, 2001: 429).

What exacerbates the general problem of homogenization, however, is the specific way in which third world women are stereotyped. In development and women's studies, 'third world women' are frequently portrayed as inherently backward, intellectually powerless, poor and generally helpless. Approached from this angle, the third world woman can only be perceived as some sort of victim of another 'average' homogenized image, "which is the third world man of globalization, universal sexism and history" (Wood, 2001: 430). These generalizations can even lend power to a stereotyped conception: that the 'typical' Western woman should always be educated, empowered and free. In development practice and academic theory, Western women are often perceived as being experts on gender and development and are also often employed as such. This results in an illusion of contrast between the stereotypes of the 'modern Western woman' and the 'helpless third world woman'; and this contrast mistakenly renders Western women the 'saviours' and third world women the 'victims.' Eventually, this results in the idea that third world women must be 'saved' by Western women, disregarding their internal needs and demands in development processes. Therefore, Western women often make the mistake of normatively presupposing that they are already aware of what is important when trying to aid and assist third world women, ignoring the diversity of personal perspectives that these women themselves have on development processes. As a result, it is often assumed that they must listen to the unheard and silenced voices of the 'third world women.' This is a rather distressing attitude since it includes a sense of superiority rather than a deeper understanding of what development for a diversity of women entails.

Examples from Practice: Female Circumcision

Taking the different approaches mentioned above in mind, it is important to shed some more light on a few different cases. To begin with, a few perceptions on the controversial issue of female circumcision will be examined. The anthropologist Christine Walley tries to find out whether the discourses and dissenting opinion in the Western world on female genital mutilation "constructs an adequate feminist and humanist political response on this issue" (Walley, 1997: 406); practices such as 'female genital mutilation' may include female circumcision, clitoridectomies, excision, infibulation, genital mutilation, while "some Anglo-Americans also observe torture" (Walley, 1997: 407). Considering the physical and emotional consequences of FGM, such practices may be perceived as subordinate, ignorant, passive and barbaric through a Western (feminist) lens, but at the same time, for women in particular African societies, who are socially forced to undergo this ritual and who approve participation in this practice, it bears an essential social function as a rite of passage for becoming a 'real woman.' Consequently, Walley tries to explore the socio-cultural setting, "in which it was desirable to 'socialize' female sexuality or fertility through female genital operation" (Walley, 1997: 407). In this way, she focuses on the motivations of these women as participants in such practices rather than dismissing them as passive victims.

A few explanations can be given for the practice of female genital mutilation: it is believed that girls bear a female and male aspect, and by removing the male aspect, they become an adult by the pain they suffer from, thus becoming a 'real woman' in the process. In addition, another motivation is that excision and infibulations also function to control their sexual desires, thus guaranteeing for a potential spouse that a woman is clean and moral before entering a marriage. In other words, this practice gives women more status in the societies where female genital mutilation is practised, but more importantly, without being circumcised, they can be even perceived as insufficient or untrustworthy. As a consequence, women can become excluded from marriage, which is experienced as a dominant social unit.

Consideration of the socio-cultural functions of such practices demonstrates a more nuanced approach. Through a Western lens, female genital mutilation is often perceived as a barbaric form of torture typical of 'backward' societies, which undermines the homogenous interests of feminism. Having said that, the physical, medical, emotional and sexual consequences for women undergoing such practises cannot be ignored when touching upon the nature/culture dualism that feminists have questioned in scientific as well as in ethnic studies. Therefore, examining this issue does not lead to a justification of such practices, but rather calls again for a deeper understanding of the socio-cultural implications it bears because it continues a traditional colonial stance by making a typical Western perception of a female body and sexuality universal in a quest to liberate women.

The Veil Issue

Another example that can be examined within this framework is the veiling as discussed by the anthropologist Saba Mahmood. Veiling is considered as essential to the ethical system Muslim women practise in the same way as praying five times a day and attending the mosque. Mahmood illustrates this in one of her ethnographies through the Mosque movement in Egypt where large groups of women call on each other "to study Islamic doctrine together in mosques and aim to educate lay Muslims in virtues, ethical capacities and forms of reasoning" (Mahmood, 2001: 203). From this perspective, the veil is regarded mainly as a sign of Islamic solidarity and submission to Allah, which manifests faith. However, seen from a Western perception, the veil is also often regarded as a symbol of oppression that imprisons the woman in a subordinate position – it is considered an obstruction of the emancipation of women. The following quote illustrates this clearly: "Women wearing the headscarf are not treated equally and cannot be treated equally, because they are veiled; for that reason, unveiling these women is seen as liberating them from the suppressive nature of the headscarf and Islam by the West" (Chaudhary, 2005: 358).

Veiled women in the Middle-East are often perceived as inferior, passive and subordinate by Western feminists, while for some women, the veil might be perceived as a symbol of self-determination and identity that indicates one's faith in Islam and a symbol of leading a moral life. Both Mahmood and Walley imply that 'the West' does lack an understanding of different socio-cultural or religious realities and should rather ask what saving is. According to them 'third world women' or veiled women, are not vulnerable because they react against the mainstream society by means of the veil as a symbol of liberation. From this point of view, Western feminists should not take their Western perceptions on gender equality and liberation as normative to societies where the earlier discussed practices of 'female genital mutilation' and also veiled women are present. Subsequently, it is important to realise that a Western lens and equality perceptions may bear entirely different meanings in other cultural contexts, where Western norms can be perceived as causes for disgrace and marginalization, and, therefore, could possibly nurture escalation of the gender inequality structures for particular women in particular societies. Seen in this light, ideologies of liberation should not be applied so easily when examining non-Western modes of human agency and, thus, should also not associate cultural and religious virtues with passivity and submissiveness, but with practices and experiences instead.

An overlapping perspective is given by the anthropologist Lila Abu-Lughod, who explores the ethics of the current 'war on terrorism' and "asks whether Anthropology – the discipline devoted to understanding and dealing with cultural difference – can provide a critical purchase on the justifications made for the American intervention in Afghanistan in terms of liberating, or 'saving' Afghan women" (Abu Lughod, 2002: 783). She elaborates on this from a politico-historical perspective and argues that

> rather than seeking to 'save' others (with the superiority it implies and the violence it would entail) the West might better think in terms of cooperating with women in need in situations that are recognized as always subject to historical transformation – while considering our own larger responsibilities to address the forms of global injustice that are powerful shapers of the world in which they find themselves (Abu Lughod, 2002: 784).

Saba Mahmood (and Abu Lughod) develop many arguments about the limits of cultural relativism through a consideration of the *burqua* and the various meanings of veiling in the Muslim world from contemporary discourses on equality, liberation and rights with earlier colonial rhetoric on Muslim women in the third world. For instance, concerning the *burqua*, she explains that "the Taliban did not invent the *burqua* since it was a local form of covering that *Pashtun* women (an ethnic group in a specific Afghan region) wore as a symbol of modesty or respectability when they went out" (Abu Lughod, 2002: 785). Contrary to what is so often argued, when it comes to the *burqua* and veil as symbols of oppression, twenty years ago the *burqua* was perceived as a 'portable seclusion,' thus, as a device of liberation, because it allowed women to travel through the public domain while still meeting the moral criteria of guarding and keeping women away from unacquainted men (Abu Lughod, 2002: 785). Seen in this light, wearing the veil and the *burqua* indicates that a woman belongs to a particular group of people and that she leads a moral life in which the family is the cornerstone of society and the home a woman's sanctuary. For that reason, it would not be appropriate to apply – from a Western normativistic lens – the Western liberation perception on Afghan women in *burquas* since what is perceived as oppressive in the West might be a liberation for Afghan or so-called third world women. Besides, from an acknowledged anthropological point of view, women are expected to respect the demanded dress codes for their social collective environments and are directed by shared standards, religious beliefs and moral values. By means of this explanation, Abu Lughod indicates the lack of understanding and reveals the politico-structural agendas based on strategic military and economic demands (such as oil interest and arms industry) that are hidden under the ruse of development. In short, in countries such as Afghanistan, where the ideology that third world women should be 'saved' is applied, feminism becomes a subterfuge for 'beneficial' politics,' since governments such as the Taliban are confused with innocent women. This results in the women being harmed instead of being 'saved.'

Ayaan Hirsi Ali and the Honour-related Violence Agenda

Another instance of negative side-effects that result from the fallacy that third world women must be 'saved' can be seen in the political work of former Dutch member of parliament Ayaan Hirsi Ali: she was received with praise by some for allegedly revealing the flaws of multiculturalism and the oppression of women within Islamic communities, even to the point of reaching the top-100 of most influential people in the world, but at the same time, she was also met with much resistance and criticism by others for her offensive use of her extremely stereotypical picture of Muslim communities as backward or even hostile. She started a public debate about women in Islam, particularly through her film *Submission* about the subordinate position of women in Islam, which followed the murder of the Dutch filmmaker Theo van Gogh by a Muslim extremist. Hirsi Ali's 'backlash' against cultural diversity in the Netherlands dovetailed with the widespread anti-multiculturalism discourse (Ghorashi, 2003). The main themes in her work are the suppression and abuse of Muslim women, who allegedly suffer from restricted freedom, domestic violence, honour-related violence, female circumcision, repudiation and forced marriage, while ignoring that at the same time, highly educated women raised in modern Western societies also suffer from some of these problems (Snel & Stock, 2006: 5). Therefore, she states that marriage, virginity and female chastity are problematic and that the Muslim lifestyle almost inevitably implies distrust of women before and after marriage. Additionally, she claims that Islam is, as a culture and religion, hostile towards women, that Muslim migrants in Western societies are traditional and should be initiated into modernity since the majority of migrants originate from the 'third world' countryside where tribal traditions still prevail with a hierarchical authoritarian attitude and patriarchal family structure. In this structure, a woman has a reproductive function and should always obey the man, otherwise she would disgrace the family – it involves a group-regulated mindset, a collective mentality featuring strong social control (Hirsi Ali, 2002: 172). According to her, in this type of culture, family honour makes people obsessed with avoiding disgrace: "in honour cultures a marriage must be entered as a virgin, otherwise women suffer from honour-related violence" (Hirsi Ali, 2003: 17).

Hirsi Ali believes that these existing local customs have been strengthened by Islam because in Muslim families there is a grave taboo against talking about contraception, abortion and sexual violence, and she implied that this taboo springs directly from Islam. According to her, children are raised in a climate of distrust, deceit and violence, that teaches girls to submit and boys to suppress them. And women lack proper education as well as the ability to provide their children a good upbringing – this results in a vicious cycle of backwardness (Hirsi Ali, 2004: 13). A second major theme in her work is her critique of Islam in general and, lastly, the inability of governmental policies to accomplish the social integration of immigrants into Dutch society because they are not strict enough. Therefore, she defines integration as a cultural problem.

Nevertheless, although Hirsi Ali has a controversial and inappropriate tone, it should be acknowledged that she successfully placed these issues on the political agenda. Therefore, it is troublesome that she takes a clash of cultures approach as the starting point of her debate. Her arguments are typical of a normative cultural discourse about 'third world women' and the 'problem' of integrating immigrants and immigration into Western societies. Moreover, her extensive media exposure exerted a great influence on Dutch public opinion ('cultural backlash'). Hirsi Ali's arguments are grounded in an outdated and problematic 'essentialist' notion of culture and religion; cultural anthropologists question whether the notion of 'culture' in these conflicts is useful at all – "they oppose the outdated culturalism in public and political debates about multiculturalism" (Snel & Stock, 2006: 12). Furthermore, Snel and Stock state, "it is not the idea that cultural issues are important for understanding contemporary multicultural issues, but culture in relationship with human behaviour 'culturalistic fallacy' since culture is not a fixed and autonomous entity but something that is constantly produced and reproduced by human agency" (2006: 9). Thus, culture cannot by itself alone explain human behaviour, nor can religion, but more importantly, culture and religion should not be confused in the way Ayaan Hirsi Ali does.

Ayaans's arguments about family, suppression of women and multiculturalism can be considered highly culturalistic since she states that "principles of Islam create the suppression and violence against women" (Hirsi Ali, 2003: 64). Besides, culturalistic thinking is generalizing as it places different communities and cultures in opposition to each other and neglects the diversity within the communities concerned. As discussed before in other cases, members of different cultures are seen as homogenous since this approach polarizes people in an 'us' and 'them' dichotomy. Ayaan justifies this by saying that the basic principles of Islam are dogmatic and therefore cannot be criticized, revised or contradicted in any way; she also states that "Muslims who migrated to West Europe took their persuasions with them and still cling to them strongly" (Hirsi Ali, 2004: 8). In the new-post migration era, Hirsi Ali considers it best for the immigrants to abandon their cultural baggage; this means that newcomers have to develop their identity as individuals and that they have to distance themselves from the traditional honour culture (quoted in Snel & Stock, 2006: 10).

Consequences of essentialist culturalistic thinking such as the approach of Hirsi Ali are that they stigmatize immigrant communities in Western societies, particularly 'third world women' communities, since Ayaan implies that immigrants are culturally inferior (Hirsi Ali, 2003: 64). She portrays Muslims as backward, unenlightened, traditional and as caught in the obsession of honour and violence against women. The consequences are that Muslim migrants tend to withdraw and focus more and more on their own identity. Consequently, instead of promoting immigrant participation into the host society, this new discourse stresses and generates the difference between migrants and the host society. Ayaan's harsh tone and critique also worked to distance her from her target group, namely, the women she allegedly fought for by putting them in an impossible position. By exhorting Muslim women to change their life radically and abandon their culture and religion, they were placed in an either/or situation, when often they were either unable or unwilling to choose between two apparently dichotomous positions.

As a result, Hirsi Ali's essentialist argumentation is highly problematic, generalizing and biased as she emphasizes boundaries between cultures and neglects diversity or cultural and religious change. This is why she fails to see how immigrants find their own way in host societies, creating new patterns that go beyond the traditional dichotomy of either assimilation or retaining cultural identity, and considers Muslim and 'third world' culture as a set of autonomous elements independent of human behaviour. From this essentialist point of view, again, religion and culture are seen as determining human behaviour while human agency is marginalized, consequently, it "bears a stigmatizing character" (Snel and Stock, 2006: 9). In short, for Hirsi Ali, fighting for women's rights does not mean fighting against Islam. Subsequently, Ayaan's culturalist discourse about the incapability of Islam and women's rights rather impedes than promotes the position of women in Muslim communities.

Internal Change and Connectedness

Considering Ayaan Hirsi Ali's approach, but also practices such as female genital mutilation and the issue of the veil as discussed above, it is vitally important that Western women who try to 'save' women in difficult positions consider what is most beneficial and necessary for the women "in need"; most importantly, they need to think beyond their own cultural framework and world of experience. Several modern critics have taken up the challenge of evaluating policies and the empirical effects of feminist development in terms that emphasize the local and situated responses of people struggling at "making a living and making it meaningful" (Bebbington, 2000; see also Fagan, 1999). From this angle, the question of how to deal with cultural others is a politico-ethical problem since it is an inner paradox and dilemma in itself for the 'West' when it comes to dealing with differences without accepting the passivity implied by the cultural relativism that is seen as an improvement of ethno-centrism, racism and cultural imperialism; furthermore, that during development processes, non-Western women should not be judged, but rather listened to. However, in order to achieve this, it is important to realize that the current development discourse contains strong Eurocentric aspects that should be done away with – a more "equal" register of solidarity must be used instead of one of "saviourism" and superiority.

Having said that, another issue here is the question of what 'saving' or 'development' for specific groups of women entails. An interesting insight is given by the development scholar Amartya Sen, who perceives development as freedom and gives a solid description of development as following: "being able to lead the kind of lives we have reason to value" (Sen, 1999: 14). This is a valuable definition of development, which raises another issue that can be clearly illustrated by a 'what you see is not always what you get' insight indicating an insider/outsider dilemma in defining the essential necessities of the women in need or in empowering them so they can strive for the lives worth living. This also reveals the complexity of getting the real picture behind the surface, since there are many factors involved in breaking from the image of 'the outsider' (e.g., by means of mastering the local language, trust-building long-term processes for grasping the complexities and sensitivities of local realities etc.). Here, it is of utmost importance to question the issue of what change is about and what the stimuli for change are beyond the surface of those who suffer from gender inequality.

In the end, women want to liberate themselves from subordinate and suppressive positions. However, one should not forget that their socio-cultural, religious, geo-political, economic and local position may be a part of their reality, and may have shaped the scope of their lives. For that reason, it is of significant importance to consider how to reach another's reality – not by stating "we are liberated and know how to liberate you" since there is no solid starting point or right to impose one set of values upon another's. This approach devalues the position of these women instead of finding identification and connectedness that covers the necessary demands from a grassroots and bottom-up approach. Seen from this perspective, it is essential to gain more insight in the 'other' world from a bottom up approach, and to inform one's experience by trying to understand the needs from 'within' the other reality in order to gain access to the underlying sensitivities and complexities. For instance, as mentioned earlier, one could ask why a veil, for one reality, is perceived as a symbol of suppression while in the other context, it symbolizes liberation. The differences in these points of view disclose the implicit expectations behind the need to listen. The enabling factor of opening the door of this 'other' reality, in order to listen within, is what creates that something new which will often appear. Having said that, it also important to mention that Western women, who believe that they are free themselves, may be imprisoned in their own reality as well. In the West, for instance, there is a forgotten 'tyranny of fashion' and 'plastic surgery industry' that could also be considered a form of oppression, a submission to the fulfilment of the idealization of physical appearance.

Top-down and Bottom-Up Interaction

Taking the aspects mentioned above into consideration, it is important to 'equalize' within the target group, since some women need more solidarity than others. One should weigh what is important in order to address the demands particularly of 'third world women' or in Western societies by 'immigrant' women themselves, as expressed in third-wave feminism or by Muslim feminists (such as the Egyptian feminist Nawal el Sadawi or the Moroccan sociologist Fatima Mernissi). Mediators, key figures or local leaders from the communities themselves should be given structural, financial as well as moral support in the same way, and be listened to so that their way of life can be understood from a diversity of perspectives. Therefore, for Western feminists, whether anthropologists or activists, it is essential to mediate with such local key persons from 'within' the targeted communities based on a bottom-up approach as they are the ones who have access to the hidden underlying sensitivities and insights such as socio-cultural codes and unwritten rules. For that reason, those key persons representing the internal needs should be supported, facilitated and guided in cooperation with top-down agencies on a local, regional, national and international level in their plea for helping women in need – they are the ones who are able to nurture self-awareness processes and gradually bring about an internal change of mentality on a long-term basis in particular societies.

In addition, also within the development literature as well as development practice, it is often claimed that top-down interventions do not always represent the internal and necessary demands of the poor. One of the major explanations for this claim lies with the lack of consideration of the complex socio-cultural dynamics that form development perceptions and programs on a local basis – these include social exclusion, manipulation and local expectations within the framework of cross-cultural communication, interpretation and translation. For this reason, it is essential to consider how re-production of 'local knowledge' representing the internal needs might be generated through focusing on such morals, standards and values of the communities themselves that are 'locally determinative' in targeted areas in developing countries or so-called 'third world' realities. As a result, it can be said that development participation requires a holistic and inter-disciplinary approach in assessing what major challenges for women in need are.

Overall, it can be said that several measures are essential for successful development: in-depth analysis by breaking through local socio-cultural dynamics, complexities and sensitivities including the voices of the underserved and disadvantaged groups and listening to their demands by providing a platform to express their needs, but also spending time with the local communities. Additionally, the importance of personal stories cannot be ignored when attempting to connect the diversity of layers within specific realities and grasping what lies 'beyond.' Besides, it is also essential that the interaction between the top-down and the bottom-up level approaches in actions and initiatives will be balanced with an individual, organizational, institutional, societal and global level. It is undeniable that the politics of participation for reaching women in need is a reality within the development practice where power relations are valued above human beings. However, empowerment to enable these women to identify and prioritize their own needs from an interdisciplinary and holistic approach in bridging both theory and practise through the engagement of all parties involved – and for the sake of providing a more equitable world, is still a responsibility for each of us that cannot be ignored.

Conclusion

In conclusion, when considering whether 'third world women' need to be saved by the 'West' while undergoing practices such as FGM, honour-related violence or veiling, the question through which lens we are looking – a Western or rather a humanitarian lens – needs to be considered; moreover, one needs to question whether that lens is normative and thus actually applicable to other realities and worlds. The practice of listening to the silenced voices of third world women must take place in a context of identification, solidarity and trust. All of this must be connected to the communities themselves, and their striving for liberation, self-determination and development in a diversity of historical socio-economic, cultural, politico-structural and demographic perspectives. In the end, the causes and solutions for the most disadvantaged women and people in general are multidimensional on a worldwide scale. It is therefore important that the task of sharing responsibility, solidarity and connectedness with the ones in need be a collective one – a task for each of us since the plea for equity for the most disadvantaged and underserved groups in societies will need a global network of dialogue and cooperation in which we will need to trust one another. Only then will it be plausible to escape from the labyrinth of clashing lenses in order to give the voiceless a well-grounded voice.

Edward Said and the Palestinian Experience

Ihab SHALBAK

PhD Candidate at the University of New South Wales, Sydney

> The starting point of critical elaboration is the consciousness of what one really is, and is "knowing thyself" as a product of the historical process to date which has deposited in you an infinity of traces, without leaving an inventory. It is important therefore to make an inventory.
>
> Antonio Gramsci, *Prison Notebooks.*[1]

By the time of his death in 2003 Edward Said was one of the most visible public intellectuals in the world. He was simultaneously admired and loathed for his academic achievements and political interventions. For his admirers, Said inspires "a rich variety of oppositional practices and committed scholarship" (Deer, Prakash and Shohat, 2006: 1). For his detractors Said represents the worst possible combination of biased, amateur, anti-Western scholarship and nativist primordial politics. Central to both views is Said's commitment to the Palestinian struggle.

Said's political and intellectual association with the question of Palestine made him a subject to various kinds of conservative and Zionist charges including labeling him a "Professor of Terror" who "spills ink to justify their [members of the Palestinian National Council] spilling of blood" (Shohat, 1995: 221). But perhaps the more serious form of charges levelled against Said were that his entire intellectual enterprise was a series of reactive and consequently reductive responses to his experience as a Palestinian. The novelist Simon Leys encapsulates this line of argument when he claims that Said's account of the imperial Western appropriation and construction of the East. Leys argues that, "*Orientalism* could obviously have been written by no one but a Pales-

1 Said used this quote by Gramsci as an epigraph to his long essay *Zionism from the Stand Point of its Victims* (1979); he repeatedly returned to this quote in his writing and interviews.

tinian scholar with a huge chip on his shoulder and a very dim understanding of the European academic tradition" (2004: 49).

In most arguments against Said there seems to be an adverse relationship between his Palestinian credentials and his intellectual stature. If being a Palestinian is not synonymous with being a terrorist, then the commitment to the Palestinian struggle is a chip that impedes intellectual development.

More recently Homi Bhabha seems to lend more elaborate support to Leys' contention. According to Bhabha, "there is much to agree with [in Said's arguments], but much to question also. The high Saidian style speaks with a moral passion that sometimes sacrifices analytic precision to polemical outrage, and his singular commitment to the Palestinian cause could create a severe hierarchy of historical choices. Said's desire to snatch some shred of dignity for his diasporic and misrepresented peoples led him, at times, to pass over distinctions, to resist shades of meaning and interpretations that might have widened the circle of empathetic dialogue" (Bhabha, 2004: 19). Bhabha goes on to claim that Said's "rage sometimes drives him toward a dark two-dimensionality" (2004: 19). Needless to say, accusations like those that call Said a 'professor of terror' are more easily refuted than the contentions put forward by a postcolonial theorist like Bhabha, that is, because the postcolonial discipline is a field many trace back to Said's *Orientalism*. Bhabha's contentions raise a number of issues in regards to the integrity of Said's intellectual enterprise. First, Bhabha gives credence to what Said's opponents have argued all along, that Said's masterpiece *Orientalism* and his subsequent works are, at best, all displacements of his intellectual attention as reaction to his Palestinian experience. Second, that there is a possible contradiction between the over involvement in politics and intellectual labor; the admired 'oppositional practices and committed scholarship' of Said hinders more thoughtful intellectual pursuits. And third, if one of the finest public intellectuals of our time degenerated into 'a dark two-dimensionality' in articulating the struggle of his people, what should we expect from the outraged people of the South, could they produce any voice or politics that does not subscribe to or reproduce binary constructions?

There is no denial that Said's intellectual and political legacy are inextricably linked to the Palestinian experience. The thesis I advance here is that far from being an impediment, Said's association, or affiliation as he would insist, with the Palestinian experience in general and the Palestinian struggle in particular was essentially an extension of criticism in which:

> critical intentionality has the capacity to negotiate between the local and the global… between theoretical insight and resistance to theory, between histo-

> ry as burden and as opportunity... In short, criticism names that which allows a given writer optimally to come to grips with distance – where 'distance' to be understood, not only geographically and culturally... but also historically, epistemologically, and politically (Hussein, 2002: 215).

Said termed his critical approach in 1984 *contrapuntal reading* according to which: "we must be able to think through and interpret together experiences that are discrepant, each with its particular agenda and pace of development, its own internal formations, its internal coherence and system of external relationships, all of them co-existing and interacting with others" (Said, 1994: 36). This comparative approach allowed him to sail against self-enclosed theoretical modes which pervaded the academic world.

Almost at the same time the emerging Palestinian resistance was moving in the opposite direction to the Arab decolonization movements, which had begun to show signs of decline, and against what seemed to be unstoppable Israeli triumphs. In the late 1960s and early 1970s the emerging Palestinian movement attempted to develop a secular understanding of the struggle against Zionism as a colonial-settler movement. The various tendencies of the movement, traditionalist, Liberalist, Islamist, Arabist, Nationalist and Marxist spent the 1960s in a state of restless intellectual and organisational flux attempting to come to terms with the Palestinian *Nakba* (catastrophe). The effort produced different factions at tangent affinities to all existing political ideologies. Rather than taking the simple commonality of language, religion and history (the mainstay of classical nationalism) the emergent movement commenced by taking the actuality of the Palestinian experience as a departure point which made possible the development of new distinctions and established affiliations with other experiences. As such the Palestinian movement was part of what Frantz Fanon described as an anti-colonial nationalism which acquires its consciousness through relating to the international dimension. For Fanon "it is national liberation which leads the nation to play its part on the stage of history. It is at the heart of national consciousness that international consciousness lives and grows" (Lazarus, 1997, 39). In the Palestinian case, the Algerian revolution, the steadfastness of Hanoi, the daring struggle of South Americans against dictatorships and so on, all provided existential and intellectual sustenance.

Traces

In 1962 the late Palestinian writer Ghassan Kanafani published a novella titled *Men in the Sun*. The novella traces the fate of three Palestinian refugees in their attempt to get smuggled to the oil rich Kuwait in order to escape the confinement of refugee camps that became permanent residences for dispossessed Palestinians after the 1948 war in which Zionist forces ethnically cleansed hundreds of thousands of Palestinians. The events of the novella takes place in an Iraqi town, where the three refugees manage to strike a deal with a truck driver of a water tanker to smuggle them in the empty belly of his tanker to Kuwait. As they approach the Kuwaiti border's checkpoint, the trio hides in the belly of the tanker while the border inspection is being negotiated. Under the blazing sun of the desert, inside the belly of the tanker, the three men slowly and silently suffocate to death.

The novella ends with the scene of the driver speeding away after he has disposed of the bodies in a landfill. At that very moment the driver utters the final sentence in the novella, he asks "*[w]hy didn't you knock on the sides [or walls] of the tanker? Why didn't you say anything? Why?*" (Kanafani, 1999: 74). In the unfolding narrative of the novella the driver asks this question to absolve himself from the responsibility of the three men's death; however, the message for the Palestinian reader of this novella revolves on the question of assuming responsibility. With this ending the work of the Palestinian reader of the novella is far from being done; rather, it is actually just begun.

In 1962 the options of the Palestinian people literally were between landfills and landmines. After 1948, the vast majority of Palestinians were deposited in refugee camps as human waste,[2] and viewed as a surplus population, that existed on the one hand as an impediment to the realization of a colonial settler society in Palestine; on the other hand as a threat to the consolidation of Arab states, particularly Western-leaning states, and consequently a threat to the Western interests within the geopolitical configuration of the Cold War. Moreover, the presence of the Palestinians, given Europe's crimes against its Jewish population, inevitably came to represent an obstacle to Europe's attempts to offshore its Jewish question particularly after the Holocaust. This compounded confluence of dynamics literally made any Palestinian attempt to step outside the landfills of the refugee camps stepping into landmines of a sort. As Edward Said noted when the three men in Kanafani's novella "finally move out of their spiritual desert into the present, toward the future which they reluctantly but necessarily choose, they will die – invisibly [and] anonymously" (Said, 2000: 52). In the 1950s and early 1960s before the emergence of the Palestinian resistance movement, thousands of individual Palestinians perished trying to return to their

---

[2] For example, one Lebanese ministry declared that "Lebanon was not a dump for human waste". See Schulz Lindholm, Helena and Hammer, Juliane, *The Palestinian Diaspora: Formation of Identities and Politics of Homeland*, New York: Routledge, 2003: 61.

homes in order to escape the political and social death they were experiencing in the refugee camps.

In his analysis of *Men in the Sun* as an example of Arabic prose fiction after 1948, Said does not offer a chronology alongside the novella to detail the Palestinian suffering, rather he offers an intricate reading of the cultural form, the novel, the history of its various incarnation and the diverse worldly experiences it articulates. Said simultaneously explores the "history in the novel and the history of the novel" (Said, 2000: 42). Said takes his reader through multiple temporalities and localities offering a dense diachronic and synchronic reading of the coming together of the Palestinian experience and novelistic techniques. The discussion of Kanafani usage of the novelistic scene demonstrates the method that Said uses to meditate on the relationship between disparate practices, imaginations and experiences:

> For Kanafani a scene is centrally the convenience given to the writer by the general novelistic tradition; what he uses in order to present the action, therefore, is a device which, displaced from the tradition that can take it for granted, ironically comments on the rudimentary struggles facing the Palestinian. He must make the present; unlike the Stendhalian or Dickensian case, the present is not an imaginative luxury but a literal existential necessity. A scene barely accommodates him. If anything, then, Kanafani's use of the scene turns it from a novelistic device which anyone can recognize into a *provocation*. The paradox of contemporaneity for the Palestinian is very sharp indeed. If the present cannot be "given" simply… it is intelligible only as *achievement*. Only if the men can manage to pull themselves out of limbo into Kuwait can they *be* in any sense more than mere biological duration. (Said, 2000: 53)

Two observations can be made here. The first relates to Said's reading of the novella and the other identifies Kanafani's usage of the novelistic techniques as observed by Said. First, Said's use of the history *of* the novel to elaborate on the history *in* the novel reveals what is lacking in understanding his intellectual elaboration of Palestine, he is often read primarily as a Palestinian with an authority to speak on Palestine and as a Palestinian with little regard for his intellectual practice; but as Tim Brennan observes,

> Edward Said's authority was always ultimately literary… His status in such varied fields as history, geography, music criticism, political commentary, expose journalism, and the Palestinian movement defers to the place of honor he precociously established within comparative literature between 1966 and 1975, and it was through literature and the tropes of literary criticism that his political and social imagination was consistently expressed during his *anni mirabili* (1975-1992) (Brennan, 2004: 23).

Even, for example, in a profoundly political and historical essay like *Zionism form the Stand Point of Its victims* (1979), Said commenced his analysis of the relationship between Zionism and imperialism by analysing George Eliot's last novel, *Daniel Deronda* (1876).

Second, the scenes in Kanafani's novellas either become or are sites of political intensity which condense and problematise the Palestinian experience, at once leading to articulate new meanings as *provocations*. Similarly, in *Return to Haifa* (1969) a novella published after the Israeli occupation of the rest of Palestine in 1967, Kanafani used novelistic techniques to provocatively interrogate the Palestinian experience after the latest Arab defeat by Israel. This is a point that is emphasised by Barbara Harlow when she observes that,

> [T]he Palestinian couple's 'return to [city of] Haifa' complicated... in Kanafani's text by flashback, multiple perspectives, stream of consciousness and other storytelling devices, is conditioned by this devastating defeat in such a way as to provide an ideological framework for the critical re-evaluation of the Palestinian vision and the means to its realization (Harlow, 1986: 5).

In *Return to Haifa*, Kanafani, by practicing journalism, literary criticism and political commentary, seems to find the most amiable form to critically undercut nostalgic narration and infuse it with the urgencies and realities of the present.

What is significant about Said's fascination with the novel form and Kanafani's utilization of it, is that, to quote Georg Lukács, the novel is "the representative art-form of our age: because the structural categories of the novel constitutively coincide with the world as it is today" (2000: 460). For Said, "The genius of this description of the novel (which is certainly the most brilliant ever offered) is that it shows how as an art form the novel reconciles within itself the internal discrepancies of modern life, and in so doing coincides 'with the world as it is today'" (Said, 2000: 460). This correspondence, which Said strongly endorses and adopts, allowed him to express and conceptualize political questions in Kanafani's novella through the tropes of literary criticism.

Kanafani, in his adaptation of the novel's techniques and categories, does not make the Palestinians a nation in-waiting based on given foundations; rather he uses the interplay between memory and the present realities to establish a new critical consciousness. For Said critical consciousness is the ability to grasp and act on the worldly conditions of one's circumstance, it "means being able – by discipline, by seeing, by a sense of community, by dedication to a set of rational libertarian value – to exist, intellectually and theoretically, in the very forms of oppression, to understand them as historical experiences, to grasp them in their cultural, material density" (Said, 1979: 53-54). The scene in the novella as site that condenses Palestinian contemporaneity seems to be about the ability to exist in the very forms of oppression in order to grasp reality.

The Palestinian reality that Kanafani expresses is the reality of being in exile, the reality of having to make your decisions, choices and priorities based on this fact. The novel form allowed him to represent this fact and dramatise it. In "Reflections on Exile", Said following Lukács suggests that:

> the novel, a literary form created out of the unreality of ambition and fantasy, is *the* form of "transcendental homelessness". Classic epics, Lukács wrote, "emanate from settled cultures in which values are clear, identities stable life unchanging. The European novel is grounded in precisely the opposite experience, that of a changing society in which an itinerant and disinherited middle-class hero or heroine seeks to construct a new world that somewhat resembles an old one left behind forever. In the epic there is no *other* world, only the finality of *this* one Odysseus returns to Ithaca after years of wandering, Achilles will die because he cannot escape his fate. The novel, however, exists because other worlds *may* exist, alternatives for bourgeois speculators, wanderers, exiles" (Said, 2000: 181-82).

In the Palestinian experience, exile both as a lived reality and an extensional situation, is not the preserve of the middle-class. Dispossessed Palestinians from all classes, not just intellectuals, faced the transformation of their existence into scattered disinherited refugees. They all had to actively search for alternatives. In the Palestinian case other worlds not merely *may* exist but they ought to exist out of necessity if not by choice. In *Men in the Sun*, according to Said, the main conflict is "about that contest in the present: impelled by exile and dislocation, the Palestinian must carve a path for himself in existence which is by no means a 'given' or stable reality for him" (2000: 52). The intellectual and ideological flux that shaped the formation of the Palestinian movement was a manifestation of this lack of any strong referential "givens" in exile. Each faction in the movement had more than one current and the main faction, Fatah, was an ensemble of all possible political views from the Maoist to the Islamist.

This significant but rarely discussed *politics of disaffiliation* (with mythologized heritage or origin) of the Palestinian movement is what drew intellectuals like Jean Genet to support it in the early 1970s. As Clare Finburgh argues Genet embraced the Palestinian fighters because their "ceaseless reinvention of politics, religion and society prevents any original cause from existing beyond the permanently unstable site of discursive relativity... [Genet's] 'Quatre heures' bears witness to the

Palestinian rejection of monolithic ideology" (Clare, 2002: 483-94). Like the novel the Palestinian movement was born in "a world that has been abandoned by God" (Said, 2000: 460). This situation forced the movement to interrogate the very categories according to which the struggle for Palestine is understood.

Ghassan Kanafani was a prime representative of this interrogative restlessness, he was a thinker and militant who moved between various ideas and conceptions in the search for a worldly answer to the challenge that Zionism represented the Palestinian people with. In his writing, as discussed earlier, the question of Palestine morphed into a series of existential and intellectual interrogations. In the 1950s, the twenty-something Kanafani started his writing career along the good/evil equation in which the Zionist represented the evil aggressor while the Palestinian was the good victim. Later Kanafani moved to write along the dichotomy of right/wrong where he relayed the moral justness of the Palestinian cause. Unfulfilled and unsettled by both formulas, from writing *Men in the Sun* until his assassination at the age of 36 in 1972, Kanafani moved to a more worldly political understanding of the Palestinian struggle centered "around the idea that self knowledge is indivisible from knowing the Other and of identifying the separate histories of both opposing parties in order to unify these histories in a new history, that is the history of the conflict that simultaneously unifies and separates them" (Darraj, 2002: 69).[3]

And this is specifically what he did in his 1968 novella *Return to Haifa* that was discussed earlier, where he tells the story of a Palestinian couple who returned to the port city of Haifa after the occupation of the rest of Palestine in 1967 to reclaim their son, Khaldun, whom they involuntarily abandoned in 1948. In Haifa the couple find that the woman who took over their house, a Jewish Holocaust survivor, has adopted and brought up their son under the Hebrew name Dov. To their dismay, the son is a soldier in the Israeli army who rejects his biological parents and only acknowledges his adopted Jewish parents.

Apart from exploring the contemporary intertwining of the Jewish and Palestinian histories in the unfolding of the conflict, Kanafani in this novella concurs that a past based on foundational conceptions, identities and filiations is beyond recuperation and unattainable, the return journey he embarked on in the novella is really an interrogation of the very categories that governed the question of Palestine/Israel. Towards the end of the novella the Palestinian protagonist asks himself "What is the

---

3 My translation from Arabic, Darraj, F., *The Memory of the Conquered: Defeat and Zionism in the Palestinian Cultural Discourse*, Casablanca: The Arabic Cultural Centre, 2001.

homeland? Is it these two chairs which have been in this room for twenty years? The table? The peacock feathers? The picture of Jerusalem on the wall? The copper door bolt? The oak tree? The balcony? What is the homeland? Khaldun? Our dreams about him? Parents? Children? What is the homeland?" (Harlow, 1986: 19). For Barbara Harlow what Kanafani did in *Return to Haifa* is that he insisted "on a re-elaborated definition of the nation, the homeland, based not, in Edward Said's terminology, on 'filiation,' but on 'affiliation' a solidarity that necessarily violates ethnic, racial, and religious boundaries" (Harlow, 1986: 19).

The task of the emergent Palestinian movement in the late 1960s and early 1970s was to find a way out of the impasse that the Zionist exclusivist project created on the ground. In search for more humane resolution to the violent collusion and enmeshing of Palestinian and Jewish histories and realities, the political and intellectual theorization of the Palestinian movement centred on developing a vision of a democratic secular state for Jews and Palestinians in the historical land of Palestine. In *Zionism from the Standpoints of its Victims* (1979), Said wrote:

> the core of what I believe is the Palestinian idea itself, which was first formulated in the 1930s by the Arab wing of the Palestinian Communist Party: the idea of a secular, democratic state in Palestine for Jews and Arabs. In its reformulation by Palestinian activists in the 1960s it has been the only political idea since Second World War to have appeared in the region with genuinely libertarian power: that it is at present the object of regretful nostalgia or contempt has prompted very few people to ask whether that is not another reflection of how unhealthy the politics of ethnocentrism, religious exclusivity, and racial purity has been. (Said, 1979: 52)

No figure embodied the ethos and practices of the Palestinian movement more than the person who introduced Said to the subject and the experience of Palestine, Ibrahim Abu-Lughod, a Palestinian intellectual and leader. He was a former Professor of Political Science at Northwestern University and later vice-president of Bir Zeit University in the West Bank. Said meet Abu-Lughod at Princeton in 1954 and together they inaugurated the Palestinian project of counter narration in the USA. In an eulogy piece titled "My Guru", Said says that "[i]n all sorts of ways, Ibrahim's rich life and his death both reflected and clarified the turbulence and suffering that have been at the core of the Palestinian experience"; he adds, "No one knew better than Ibrahim how to turn the shambles of defeat into some sort of achievement" (Said, 2001: 19-20). Unable to return to his native city, Jaffa, which he was forced to leave in 1948, Abu-Lughod settled in the West Bank in the late 1990s after the Oslo agreement. His entry to the West Bank did not mean that he approved of the exclusionary terms of the agreement which effectively consigned Palestinians into tiny disconnected areas in order to separate them from Jewish Israelis and at the same time denied the rest of Palestinians their right of return in order to guarantee a Jewish majority in the state of Israel. Remaining loyal to the early reformulation of a non-exclusivist state in Palestine, Abu-Lughod staged his journey back to Jaffa by using the privilege of his American citizenship to force the Israeli authority to allow his body to be buried in his native town. Far from being an epical act of finality, Abu-Lughod reopened the question of Israel/Palestine and reemphasised the essence of the Palestinian experience. In Said's view:

> [d]espite that theatrical conclusion, a vast instability remained. He was still unfulfilled and unsettled. The return didn't change him [to the West Bank], though he was more contented at home than he had been in exile. For him, Palestine was an interrogation that is never answered completely – or even articulated adequately. Everything in his personality confirmed that restlessness... His life... was a version of Palestine lived in all its complexity. (Said, 2001: 19-20).

The significance of the intellectual effort led by Said and Abu-Lughod in the USA, according to Ella Shohat, is that it "has challenged Zionist self-aggrandizing ideas of exile, diaspora, and return" (Shohat, 1995: 222). By interrogating the very categories that shaped both the Jewish and Palestinian experiences, rereading them as Kanafani has done, in relation to each other and in relation to their association with the histories of the victims and oppressors, Said and Abu-Lughod had undermined the Zionist ability to monopolize these notions in foregrounding an exclusivist and ahistorical imagery of the nation. Both Abu-Lughod and Said (the first in his death, and the second almost to the last moment of his life) mounted the ultimate challenge to Zionism by insisting on a nonexclusive, secular, democratic Jewish/Palestinian state.

Inventory

The Palestinian idea for Said represents a commitment to a libertarian project rather than allegiance to ideological prescription; as practice and intellectual elaboration it belongs to the transitive tradition that Said himself embraced in his 1972 book *Beginnings: Intention and Method.* In *Beginnings*, Said argues that the transitive intellectual approach commences by posing questions about concrete problems rather than relying on ready abstractions. For him this approach is

> a problem- or project-directed vision of beginnings aimed at enabling transitive point of departure. Transitive because it is temporal... it foresees a continuity that flows from it. This kind of beginning is suited for work, for polemic, for discovery. It is what... allows us to initiate, to direct, to measure

> time, to construct work, to discover, to produce knowledge (Hussein, 2002: 80).

Contrary to the transitive approach, the intransitive approach lends itself more easily to ideological construction. By focusing on beginning as origin, it imposes a strict identity, essentialist point of origin, on social reality. As Abdirahman A. Hussein notes, the implication of intransitive line of thought

> is that, whatever its ultimate ontological status or causative power an intransitive beginning amounts to an epistemological mystification, an "unknown" – and ultimately unknowable –'event' which cannot be operationalized inside history. It is for all practical purposes indistinguishable from silent origin… intransitive beginnings are self-subsistent, self refereeing, closed in upon themselves (2002: 80).

Joseph Massad argues that for Said, "Palestine *became* a point of departure, or more precisely a 'beginning' as opposed to an origin, a notion he associated with the theological, which had little purchase in his decidedly secular life" (2008: 123-24). Said says something to that effect in a poetic recollection of a conversation between himself and the late Palestine national poet Mahmoud Darwish. In the conversation, Said views his engagement with the Palestinian identity itself as a tool of resistance and an ethical undertaking, not an origin or finality:

> What about identity? I asked.
> He said: It's self-defence…
> Identity is the child of birth, but
> at the end, it's self-invention, and not
> an inheritance of the past. I am multiple…
> Within me an ever new exterior. And
> I belong to the question of the victim. Were I not
> from there, I would have trained my heart
> to nurture there deers of metaphor…
> So carry your homeland wherever you go… (Mahmoud, 2004).

Palestine as a departure as opposed to an originary point is indeed a defining moment in Said's political and intellectual biography. Establishing this actuality allows for an understanding of Said's complex and fruitful relation to Palestine and its relevance to his sustained critique of the cultural and political constructions of imperialism. Said's intellectual role in relation to Palestine can be divided into two parts: first, in terms of what historical role was available to him as an intellectual born into a vanishing nation after 1948, and later what historical role was available to him as an intellectual of a people to come.

As an intellectual who was born to a vanishing nation, Said's work until the 1960s was largely apolitical as he had not consciously reflected on his fate as a member of dispossessed people. As Said himself says, "I was born in Jerusalem, so was my father, his father, and so on; my mother was born in Nazareth. These facts were rarely mentioned. I earned my degrees, I became a professor, I wrote books and articles about European literature… [And] in 1967 I was 'from' Lebanon" (Said, 2000: 24). At this early stage Said, despite his elite education, did not grasp the impulse or own the language to express and articulate the Palestinian experience nor did he feel the urge to construct an originary identity of some sort.

Indeed, when Said wrote his early thesis on Joseph Conrad he was detached from the question of Palestine; he was completely immersed as a young academic in the literary scene of the elite American universities. Nonetheless, it could be argued that the intellectual enterprise that Said embarked upon during his apolitical period was imbued with the quest of deciphering fundamental issues like the construction of human reality, through institutions, will and desire as exemplified by Conrad's life and career. Said says of his interest in Conrad's legacy "Over the years… I have found myself writing about Conrad like a *cantus firmus*, a steady groundbass to much of what I have experienced" (Said, 2000: 3). He goes on to say that "Conrad's predicament was that… he was a man of action urgently in need of a role to play so that he could locate himself solidly in existence" (Said, 2000: 12). This correspondence between the exilic experience and transitive acts of Conrad and the Palestinian situation seems to allow Said, years later, to deduce from the object of his early academic study a realization of one of his true subjects, that is, the role of the intellectual in political struggles.

It was the war of 1967 and the emergence of the Palestinian movement that made the young academic rethink his intellectual position and role. As he says, "For the first time since I had left to come to the United States, I was emotionally reclaimed by the Arab world in general and Palestine in particular" (Said, 2000: 14). And here his role as an intellectual of a people to come commenced in earnest.

In his first major political article on the question of Palestine titled "The Palestinian Experience", Said observes first hand the liberationist imperatives of the emerging Palestinian movement; he says,

> In Amman today two ways of life enclose all the other ways… These two are being a refugee in a camp and being an active member of one of the resistance groups… [for Palestinians] Amman has become a terminal with no *raison d'être* than temporally to preserve displacement; beyond the city, physically and in consciousness, are desert and extinction. In Amman the Palestinian either stays on as best he can, or he repatriates himself from it as a guerrilla. He has really stopped thinking about Kuwait, or Beirut, or Cairo.

> He has only himself to consider now, and what he discovers, by what technique he uses, is how he is a Palestinian – or rather, how he has already become a Palestinian again and what this must mean to him... What has emerged, in short, is Palestinianism (Said, 2000: 21).

Said defines Palestinianism as "a political movement that is being built out of a reassertion of Palestine's multiracial and mulitreligious history. The aim of Palestinianism is the full integration of the Arab Palestinian with land, more importantly with the political processes that for twenty-one years have either systematically excluded him or made him a more and more intractable prisoner" (2000: 16). As we can see again the Palestinian identity that Said engaged with here is not an identity of recovery, in nationalistic nostalgic sense; rather, it's an identity of discovery, a self-knowledge, that comes with a conscious transitive will to transcend the physical and mental entrapments of the moment, be it a refugee camp, a water tanker, or an Arab city.

This non-identitarian politics is what placed the Palestinian movement in the 1960s and 1970s as a radical temporal or worldly, as opposed to timeless and theological, political response to the construction of an identitarian exclusive nation-state in historical Palestine. And this is the main line of affinity between Said and the Palestinian movement. It coincides with Said's articulation of the distinction between the theological politics of filiation and the secular politics of affiliation. For Said as David Lloyd explains: "Filiation respects the patrimonial dimension of culture as inheritance or legacy – it is virtually a matter of genealogy or descent, culture becoming a more or less unreflective assumption of a 'natal' belonging. Affiliation, on the contrary, appears as a reflective and chosen relation to the cultural inheritance, one that demands and founds a critical engagement" (2009).

The points of affinities I have just outlined are the stable groundbass of affiliation between Said and the Palestinian people and their movement. This is, in my view, how Said came to identify with the question of Palestine from two intertwined stand points: first as a critical intellectual who refuses dogmatic filiations and, second, from his existential personal position as exiled Palestinian. The synergy of the two positions gave us the public intellectual that we have come to know as the eloquent advocate of Palestinian rights in the West.

The Construction of the Colonial State in Sudan

Tribe, Region, Race and Colonial Power in Darfur

Noah BASSIL

Lecturer, PhD, and Deputy Director of the Macquarie University Centre for Middle East and North African Studies, Sydney

Introduction

In 2004, the previously little known region of Darfur in the western Sudan was catapulted into the international spotlight. A major conflict had broken out leaving thousands dead and leading to hundreds of thousands of refugees fleeing the violence. The images and commentary that subsequently emerged portrayed the humanitarian crisis that engulfed Darfur as a product of a tribal conflict between Arabs and Africans with the Sudanese government supporting the Arab tribes. Attached to this simplistic notion of identity conflict in Darfur was the assumption that these tribal antagonisms were age-old and represented a historical fault-line between distinct Arab and African civilizations. That view borrowed heavily from representations of the fifty-year conflict between northern Arab Muslim Sudan and southern African non-Muslim Sudan. While some similarities exist between that north-south conflict and the more recent conflict in Darfur, the substantial differences were often overlooked. As a consequence there was a certain confusion surrounding events in Darfur as many initial commentaries incorrectly located Darfur in the southern Sudan or misrepresented Darfur as a region inhabited by non-Muslims. Eventually, the information about Darfur improved. However, despite increasing political, historical, geographical and cultural examination of Darfur by the western media, academics and policy-makers, a dominant discourse regarding Darfur has emerged which regards the conflict as a result of primordial racial and tribal antagonisms.

What this chapter aims to do is historicize the tribal, racial and regional aspects of the contemporary conflict in Darfur by exploring the profound changes that Sudan underwent during the fifty years of colonial rule. There is little dispute in the specialized academic literature (de

Waal, 2005; Daly, 2007; Mamdani, 2009) that the conflict in Darfur has tribal, regional and racial dimensions. However, there is also a general consensus in that same academic literature that the conflict is not racial or tribal in terms of the underlying causes of the conflict but that the conflict can be traced back to the dynamics of post-colonial Sudanese politics. While debates over the causes of the conflict continue in the academic arena, there is little nuanced understanding of the complexities of identity formation in Darfur or of the impact of the colonial period on the tribal, racial and regional differences that have emerged to shape Sudanese post-colonial political relations. This chapter does not aim to explain the causes of the conflict in Darfur but only to contribute to an understanding of the profound impact of the colonial period as a departure point for understanding more recent events.

The "Thin White Line" of British Rule in the Anglo-Egyptian Condominium

The basic logic of British colonialism can be said to have been based on three premises. Firstly, colonial governments were concerned with the consolidation and maintenance of effective control over the "natives". Secondly, London had made it clear that it was required that all colonies were self-financing and thirdly, there was the broader imperative for creating colonial economies that served the interests of the Empire. In governing the Sudan according to these basic principles of colonialism, the administrators grappled with the contradictory effects of their policies, the peculiarities of a vast and diverse colony, and the problematics of ruling over a foreign people in a foreign land. The colonial state that developed did so as a consequence of the application of the colonial logic to concrete realities of what was encountered in the Sudan.

Colonialism in Sudan, as elsewhere, was sustainable, most of all, due to the unerring faith in the "white man's burden" and the certainty that western Europeans had in the power they held over the colonized, as Nicholas Dirks explains:

> Colonial conquest was not only the result of the power of superior arms and military organisation as important as these things were. Colonialism was made possible, and then sustained and strengthened as much by the cultural technologies of rule as it was by the more obvious and brutal modes of conquest that first established control on foreign shores. (Dirks quoted in Cohen, 1996: 42)

The "cultural technologies" first and foremost rested on the power of "discovering" and ordering the natural and human worlds inhabited by the colonized. In particular, power came from the capacity to decide on, and then order, the identities of the ruled. African and Arab, northerner and southerner, semi-civilized and uncivilized, rural and urban, and European and non-European were all classifications animated by the colonial government in Sudan, and each set of identities has left behind an indelible mark. Race and tribe, in particular, were the most powerful means of control in the Sudanese context. Sudanese identities in the colonial period were shaped by the exigencies of controlling and administering a huge territory comprising diverse social and political communities, faced with serious economic limitations, and the byzantine politics of the Anglo-Egyptian relationship.

The British anxiety over the role of the Egyptians in the Sudan led to a constant pursuit of ways to reduce the Egyptian presence. This aim was continually hampered by the unavailability of British manpower (colonial administration was restricted to males only) in the ever-expanding Empire of the late 19th century and early 20th century. In the first years after the conquest of Sudan, British personnel were reassigned to southern Africa to deal with the Boer uprising, depriving the colonial government, in Sudan, of much needed military and administrative personnel (Daly, 1986: 82). A decade of consolidating the colonial system was then interrupted by the outbreak of the First World War and the redirection of British resources to fighting the war in Europe. In this period, covering the first two decades of colonial government in Sudan, Egyptian assistance was essential for ruling a vast, and in certain places, unreceptive land (Beshir, 1974: 25-25). Nevertheless, even with Egyptian participation and the increasing levels of indigenous collaboration with the new regime, the notion of an effective Anglo-Egyptian state was fictional in most of the country. The Condominium government had control over Khartoum and the regional capitals, including El Fasher in Darfur, but had little reach beyond those limited spheres of control. In Darfur and the southern Sudan, rural based rebellions against the Condominium government continued into the 1920s. Thus, colonial rule over Sudan was tenuous in some areas and non-existent in others for at least the first two decades of the 20th century. Another factor which inhibited the creation of a more comprehensive form of colonial rule in Sudan, was the widely held British view that the Sudanese were unsuitable for the task of governing their own affairs. This view would slowly change with time.

The solution to the Egyptian problem was laid out by the British Consul-General of Egypt Lord Cromer (1883-1907) when he argued that the answer to Sudan's administrative deficit was in training "a number of young English civilians" who would come to regard Sudan as a lifelong vocation (Daly, 2006: 83). While the outbreak of the First World War intervened to delay the implementation of Cromer's plan, in time the Sudan Political Service would come to be regarded within Britain as "the most admired body of British administrators in Africa... second only to the Indian Civil Service world wide" (Woodward, 1985: 40). Although the Sudan Political Service would expand over the course of time, the numbers of trained administrators would always remain too few to administer the entire country effectively. Thus, controlling one of the largest territories in the British Empire required more than just the creation of a professional body of administrators if the British were going to be able to fill the void that would be left behind if the Egyptians were expelled from the Sudan. As anti-British sentiment in Egypt intensified at the end of the First World War along with signs of an increase in Sudanese based pro-Egyptian sympathies, British motivations for excluding the Egyptians from the Sudan heightened. However, the British were also sensitive to the fact that any decision to expel the Egyptians from the Sudan could potentially provide the spark to light the tinderbox of anti-British feelings in the region. Thus, the British required a *cassus belli* and legitimate grounds on which to curtail the Egyptian involvement without leading to massive Sudanese and Egyptian backlash (Warburg, 1971: 23). Before long the *cassus belli*, as fate would have it, became available, but the question of finding adequate manpower to rule across the country remained just as problematic. However, a program which had been used effectively in India in the 19th century, and later in Nigeria, had begun to be trialled in Sudan and at this crucial moment seemed to provide the answer which the colonial administrators had been searching.

Indirect rule had been framed in practice by Frederick Lugard in his time, first as High Commissioner in Northern Nigeria from 1900 to 1906 and then as the Governor of Nigeria from 1912 until 1918. In 1922, Lugard explained the merits of his system of indirect rule in his influential treatise on colonialism in Africa: *The Dual Mandate in Tropical Africa.* Lugard explained that indirect rule was a system of government that provided the means by which colonial governments could protect the integrity of indigenous cultures and demonstrate sensitivity to the traditions of subject peoples while allowing limited self-determination. The dual mandate, he argued, was based on the premise that European colonialism would serve to gradually introduce the natives to the civilising influences of modern government and to the modern economy, both of which had to remain under the control of colonial administrators until the natives were sufficiently qualified to accept the responsibilities of self-government. In the words of Peter Woodward, dual mandate was British pragmatism "backed by a derived ideology... of serving both Africans and the world" (Woodward, 1989:

4). Essentially, the dual mandate was a useful justification for colonialism and the forced extraction of labour and goods from the colonies.

Initially, the introduction of indirect rule to the Sudan in 1922 was limited to a few of the most geographically isolated regions including Dar Masalit in Darfur's far western reaches (Daly, 2007: 127-129). A new urgency to implement indirect rule more widely across the Sudan presented itself with the assassination of the Governor-General Sir Lee Stack in Cairo on 19 November, 1924 (Woodward, 1986: 18). In the wake of the assassination, the British government issued an ultimatum that all Egyptian personnel from the Sudan be immediately recalled. As the Egyptian troops were preparing to evacuate, three platoons of Sudanese soldiers and four officers mutinied and engaged British troops in a night-long battle (Abdin, 1985: 92-93). The Sudanese troops were defeated and the Egyptians subsequently left. British intelligence blamed Egypt for inciting the unrest and this only reinforced the integrity of the decision to terminate the Egyptian involvement in Sudan (El-Amin, 1992: 177). The conundrum now facing the British in Sudan was how to administer the vast territories under their control without Egyptian assistance and with too few British administrators at the disposal of the Sudan Political Service. The response to the problem of the weakness of the colonial façade, or the "thin white line", as Anthony Kirk-Greene (1980) has described the thin layer of British administrators that governed large colonies, was to impose indirect rule as an administrative system across all of rural Sudan, which included Darfur.

In Sudan, in the wake of the expulsion of Egyptians in 1924, the colonial government's attitude to the Sudanese also changed to meet the new requirements. In rural Sudan, traditional leaders became necessary agents of stability. Also, tradition was seen as a way of insulating the natives from the unsettling affects of social change and, importantly, would "protect" Sudanese from the corrupting influence of Egyptian nationalism from north of the border. Additionally, not all Sudanese were suitable for the task of government as C. P. Browne makes clear when he describes the urban elite as "the irresponsible body of half educated officials, students, and town riff-raff"; instead he advised that "the solid elements in the country, sheikhs, merchants" were the most trustworthy individuals to be invested with "magisterial powers" (Stiansen and Kevane, 1998: 125). Despite, the rhetoric of the "civilising mission", stability and not progress, was the cornerstone of the colonial system, and indirect rule was widely perceived as the most effective way to maintain stability across rural Sudan.

As the policy of indirect rule was implemented in the rural areas of Sudan, colonial rule which in practice came to resemble a form of dual authority rather than a dual mandate, in the sense that urbanized Sudanese were governed differently from their rural compatriots (Mamdani, 1996). In particular, indirect rule was implemented in the western and southern Sudan where geographic distance and the backwardness of the natives were perceived by the British to create the most difficulties for efficient administration. While the British considered the south the most backward region of the Sudan, it was in the western Sudanese regions of Kordofan and Darfur that indirect rule was most rigorously implemented. It was not the environment that determined that indirect rule was widely instituted in Darfur and Kordofan before it was extended to other parts of the Sudan. Rather, it was due to the pre-existing knowledge of the two western regions collected by the architect of indirect rule, H.A. MacMichael, who was Civil Secretary of the Sudan, that determined that Governor-General Sir John Maffey's order for the widest possible application of indirect rule be concentrated in the western Sudan (Mamdani, 2009).

In the end, indirect rule replaced the limited efforts undertaken by the Condominium government to construct a civil authority for administering Sudan. The enacting of the Native Administration Ordinance (1927) extended to the Southern Sudan, Kordofan, Kassala, Nuba and the province of Darfur. However, as Alex de Waal points out, the recourse to traditional structures of rule in Darfur, except possibly in the Sultanate of Dar Maasalit, was impossible in a region of exceptional ethnic complexity and transversal. Rather, indirect rule in Darfur, "was the creation of a new hierarchy of tribal administrators" aimed at regulating and regularizing tribal identities and boundaries in Darfur (de Waal, 2004). While indirect rule differed from locality to locality, the policies of indirect rule were framed by the same exigencies of parsimony, social stability, and prejudice as elsewhere in Africa, and as such the legacies of indirect rule that have been the focus of an increasing amount of analysis of the character of tribal and ethnic identities in Africa has implications for understanding identities in Darfur.

Native Administration, Tribalism and the Construction of Ethnic Identity in Darfur

The devolvement of power to local political structures that occurred during the forty years of colonial rule is an important element in understanding the solidification of distinct tribal entities in the region of Darfur. Native administration, as the policy of indirect rule came to be known in the Sudan, not only segmented and classified communities based on the tribal ethnicities that were presumed to exist but invested power with traditional *shaykhs*, who were often elevated to occupy roles as traditional rulers without evidence that such a tradition existed. As

Terence Ranger has shown, much that has been considered traditional in Africa was actually an invention of colonialism (Ranger, 1992).

As noted earlier, the colonial administration in Sudan was a "thin white line" but in Darfur the line was thinner than elsewhere in the Sudan. The conquest of the Sultanate at the height of the First World War when the British were stretched by the war effort, determined there would be few British soldiers or colonial administrators available to govern in Darfur. In fact, the British presence was limited to only four British colonial administrators in the first six years of colonial rule over the region. Under pressure to consolidate colonial control over a vast region with very limited human and financial resources. H.A. MacMichael recommended, at the time of conquest in 1916, that the administrative structure that had been employed by the Sultanate of Darfur remain essentially unchanged.

The old regime was preserved, except that the *mudir* (governor) Colonel P.J.V. Kelly replaced the last sultan 'Ali Dinar, but otherwise much remained the same. The expedience of such a policy was illustrated by the decision taken by Kelly to place his office in the former sultan's throne-room. Daly (2007: 118) provides a thoughtful description of MacMichael's attention to retaining as much of the sultanate's structure as possible. Daly believes that it was this factor which accounted for the relatively peaceful transition that took place in 1916-17, as 'Ali Dinar's rule was replaced by the Anglo-Egyptian Condominium. Daly's examination implies that the conquest of Sudan and the incorporation of the Sultanate into the Condominium changed little in the region, well at least until native administration was implemented. On a superficial level, native administration seemed to preserve the old order as the ruling families that formed the aristocracy in the time of the Sultanate maintained their positions in the new order implemented by the colonial government. While native administration was a form of benign neglect, as M.W. Daly argues (2007: 125), and colonial involvement in micro-level administration across Darfur was minimal, native administration left a lasting legacy on the region as ethnicity and ethnic boundaries hardened in a way that was a clear break from the Sultanate's policies of de-ethnicizing the region.

Native administration was partially instituted with the enacting of Powers of Nomad Sheikh Ordinance of 1922 which applied to approximately three hundred *shaykhs* across the Sudan. By 1923, the implementation of the policy had led to the ordering and regularizing of the traditional judicial functions of nomadic *shaykhs*. In 1927 the Powers of Sheikhs Ordinance further extended the powers and authority of the *shaykhs* of nomadic groups to many of the sedentary communities across the Sudan. In the western provinces of Sudan, authority over land

and legal jurisdiction, including the power of punishment and incarceration, and the power to issue fines, were invested in *shaykhs* (Asad, 1970: 168-176). In 1928, the legislation was amended to extend the legal powers of *shaykhs* by investing them with the power to preside over legal matters pertaining to each particular tribe in tribally specific native courts. Land allocation and the settling of disputes over land ownership and usage were also regarded as tribe specific and as a result the territory over which a tribal *shaykh* could regulate was demarcated. Many tribes were allocated clearly defined tribal homes, or *dars*, which made self-contained and separate tribal identities and boundaries important features of the policy of native administration. Native administration, thus, overturned a two or three hundred year process of centralizing power in the hands of the Sultan's administration and undermined a process of constructing a political and social system in Darfur where primary loyalties were owed to the Sultan, and not to tribal *shaykhs* or ethnic leaders. Yousif Takana in his study of conflict in southern Darfur has commented on the colonial policy of tribal delineation:

> The third device of the colonial administration was the demarcation of the tribal Dars. Looking at the colonial literature concerning Darfur since the inception of so-called native administration in the 1920s, when the whole of Darfur province was divided into five districts and one commissionerate, one is struck by the huge efforts that colonial officers put into the demarcation of tribal boundaries. The final outcome of this process of tribal lands demarcation in the late 1920s was a rigid tribal boundary with detailed maps fixing tribal identities in complete isolation from each other. Again, this policy of "tribal fixation" within recognized territorial lands created conflict-prone ethnic communities. (Takana, 2008)

MacMichael's own study of the Arab tribes of Sudan was premised on the underlying belief in the existence of self-contained and identifiable tribal groups. Even as MacMichael asserted the hybridity of the "tribes" that he classified, he assigned ethnic and racial labels with a certainty very few anthropologists would be prepared to do today. But tribe and race were everything to the British in the colonial period, as they mapped the Empire in terms of a human geography very much invented to meet the expectations of administration and prejudice alike (Mudimbe, 1988, Apter, 1999).

The colonial government was adept at fragmenting the region by emphasising tribal differences as a way of separating one social group from another. By extending the authority of a *shaykh* to apply where another's authority was inadmissible, the colonial rulers were able to create tribal loyalties which were then reinforced by land grants to each tribe, as designated, by the regional governor. Summarizing R.S. O'Fahey's (1980 and 2008) seminal studies on the Keira Sultanate it is clear that the colonial order differed in significant ways from past practices. In the pre-colonial period, the core of Darfuri society was held together by a common allegiance to the Sultan who ruled with the authority of Islamic tradition and the legitimacy entitled to a descendant of the Prophet combined with the customary symbolism and power of an African King. Furthermore, the Sultan retained the ownership of land and absolute power to allocate land or deprive individuals or groups of land as he saw fit. The Sultan allocated land based on loyalty and service to the kingdom and expected from his loyal retainers an acceptance of the cultural and ethnic markers that characterised the Sultanate. De Waal has described the dominant identity of the Keira as "ethno-political" which he argues changed from "a process tightly focused on the Fur identity (from about 1600 to the later 1700s), to a more secular process in which the state lost its ideologically ethnic character, and ruled through an administrative hierarchy (up to 1916)" (de Waal, 2005: 1-2).

Taking such a perspective does not preclude the existence of group differences. The key point is that ethnic identities co-existed within a political system comfortable with assimilating and incorporating ethnic heterogeneity. The Sultanate of Darfur was involved in constructing a form of Darfuri nationalism and had produced a dominant Fur identity which was neither ethnic or tribal, as often represented, but a broader and more inclusive identity based on membership of the Sultanate, and loyalty to the state.

The British conquest of Darfur established an administration in the region that differed philosophically from that of the Keira Sultanate. Where the Sultanate emphasised unity and integration, British administrators, in the words of Sir James Currie, searched enthusiastically for "lost tribes and vanished chiefs" (Holt and Daly, 1988: 137). Native administration, Daly describes as "an anomalous attempt to 'tribalize' people who had no memory of tribal authority or desire to recall it" (Daly, 1986: 367). The tribalisation of the people of the Darfur region created a terrain where self-contained and distinctive group identities developed. The policy of indirect rule resulted in a decentralised power structure under the authority of *shaykhs*, *nazirs*, *maqdums*, and *omdas*, where tribes became the primary unit of social organisation. The benefits bestowed on tribes, including recognition of semi-autonomy and an allocation of land, or a tribal *dar*, left those groups that failed to be recognized as tribes in their own right, as outsiders in a system dependent on tribal identity and fixed tribal homelands.

Despite the best efforts of the colonizers to impose order through the tribal structure, cooperation, interdependence and integration between the various social groups in Darfur continued to complicate the ethnic map and relations of the region. However, tribes had been invented and tribal lands allocated on the basis that certain areas of Darfur could be identified as the traditional lands of particular ethnic groups. In doing so, the colonial logic of ordering the natives overturned the pre-colonial Darfuri logic of interdependence and unity which had been responsible for the development of an increasing homogenization of the communities of Darfur under the rule of the Sultanate.

The British Policy of Segregating Darfur

In 1922, the *Closed Ordinances Act* followed by the *Permits to Trade Ordinance* of 1925, were enacted by the Condominium Government with the aim of preventing Khartoum's emerging political and economic elite, also referred to as the riverain Arabs, from infiltrating rural Sudan and upsetting the tenuous order that had been established in Sudan's peripheries. In particular, it was intended to segregate the southern Sudan from the north. The expanding cultural and economic influences of the riverain Arabs in the south were considered to be a danger to the British hold on the country. This fear of the spread of Arabism to the south is evident from the short communiqué issued by the District Commissioner of Raja in 1935:

> I notice that in spite of frequent requests to the contrary, large quantities of "Arab" clothing are still being made and sold. Please note that, in future, it is FORBIDDEN to make or sell such clothes… No more Arab clothing is to be made from today: you are given till the end of February to dispose of your present stock. (D.C. Raja quoted in 'Abdel Rahim, 1969: 78)

Trade relations, in particular the role of merchants, had always been a conduit for the expansion of Arabic language and culture, and of Islam in Africa. The expansion of Arabism through dress, language and the spread of the Islamic religion threatened to undermine British colonial dominance of the south. British economic imperatives were thinly veiled behind the politico-cultural project of entrenching traditional cultures in southern Sudan through indirect rule and segregation as a means of control. In this way, the British were able to eradicate economic competition in southern Sudan while simultaneously preventing the expansion of Islam. Control of the economic relations was a vital element of colonial power (Amin, 1972).

In addition, the *Closed Ordinances Act* (and the *Permit to Trade Ordinance*) was aimed at also inhibiting interaction between the riverain Arabs and Darfur's "natives" (Abdel Rahim, 1969: 75-76). However, Arabism and Islam were already well established in the western Sudan and the *Closed Ordinances Act* could not have had entirely the same intentions as for the southern Sudan. In particular, Darfur represented to the British the constant fear of resurgent Mahdism (Woodward, 1985: 43). In 1882, the Mahdist revolt had emerged from the western Sudan and within three years had swept away the first era of Egyptian colonialism in Sudan, culminating in the siege of Khartoum and the death of General Charles Gordon. The Mahdist revolt and the legend of Gordon were recurring themes which were replayed over and again as justifications for the British conquest of Sudan and the consolidation of colonial rule over the country. Hence, every British colonial officer engaged by the Sudan Political Service, from the most junior to the governor-general, was familiar with the threat of western Sudanese fanaticism in the form of Mahdism. The ever-present threat of Mahdism proved an important influence on colonial policy in Sudan.

The danger that Darfur was being heavily influenced by Mahdism emerged in the same year that the *Closed Ordinances Act* came into effect when British officials noticed the flow of people from Darfur to the spiritual home of Mahdism, located at Aba Island south of Khartoum, and then back to Darfur again (Woodward, 1985: 42). Sayyid 'Abd al-Rahman, the posthumous son of the Mahdi, had survived the initial hostility of Kitchener and by the time of the outbreak of First World War had shown his usefulness to the British authorities. The British welcomed 'Abd al-Rahman's loyalty to the Government, but always with one eye open to the possibility of his betrayal. With this in mind, the authorities used the *Closed Ordinances Act* from 1922 onwards to restrict the 'Abd al-Rahman and his agents access to Darfur. There can be little doubt that the combination of Darfur's long border and the meagre resources of the Condominium failed to completely seal Darfur from external influences, including those from 'Abd al-Rahman. Even if the *Closed Ordinances Act* could not have achieved the Sudanese government's aim, as Sudanese scholar Muddathir 'Abdel Rahim writes, of making each province as "self-contained and independent as possible" (1969: 73), the segregation policy had a detrimental impact on Darfur in a number of ways.

The first of the damaging impacts of the segregation policy was the way that Darfur was disconnected from the links the Sultanate had cultivated with the *dar-al-Islam*, or Islamic world of which Darfur had been increasingly integrated since the late 1700s at least. The recurring representation in the popular press since the outbreak of the crisis in the region in 2003 that Darfur was at the end of the earth, a literal backwater, is highly dubious to say the least. Darfur had existed at the crossroads of Africa for centuries prior to colonial conquest, and over time, the Islamization of Darfur had opened possibilities for Darfur's elite to experience the education, trade, and religious offerings of Islam. In the 19th century, the Sultan sent his representatives to study at the Al-Azhar

University, Cairo's prestigious Islamic university (O'Fahey, 1997, Reichmuth, 2004). Trade relations with Egypt and with the Maghreb had continued throughout the 18th and 19th centuries ensuring that the Keira Sultanate was integrated into the wider Islamic setting as both a provider and recipient of goods. The *dar al-arba'in*, or forty days road, from Darfur to Egypt was one of the most famous in Africa, so much so, that during Bonaparte's ill-fated attempt to conquer Egypt in 1798, Bonaparte sent a message to the Sultanate of Darfur requesting a large shipment of slaves to assist in rebuilding his army.

There is little doubt that throughout the history of the Keira Sultanate Darfur had been connected with, and integrated within, the wider Islamic world. Colonial rule, while never being able to completely erase such a history or hermetically seal Darfur from the world, was responsible for diminishing Darfur's association with its traditional relations in North and West Africa. The principal affect was that Darfur no longer had the formal institutions by which it could maintain the official diplomatic contacts that had been established with the Muslim states of North and West Africa during the era of the Sultanate. Individuals in Darfur could continue to interact with the outside world but Darfur as a whole was forced to look eastward to the Sudanese capital at Khartoum. However, Khartoum had little interest in Darfur and this was shown not only in the segregation policy just explained, but also in the general ambivalence of Sudan's colonial government towards the western Sudan.

Education policy was one area where Darfuri suffered from the prejudices of the colonial government. The colonial government actively restricted education in Darfur. It was generally believed by colonial administrators in Africa that anything more than the most practical education "posed a greater threat to the country than no education at all" (Holt and Daly, 1988: 124). Native administration was based on creating an administrative model that promoted stability, and any education policy that might lead to social mobility was not only deemed dangerous, but contradictory to the very philosophy of native administration and colonialism, despite the constant assertions of the colonial powers, colonial officials, and the ideologues of colonialism such as Frederick Lugard and Margery Perham that colonialism was first and foremost planned to "civilize" the colonized.

It was clear from the establishment of Condominium rule in Darfur that the authorities were unwilling to permit the "evil" of modern ideas to infiltrate the region and spoil their chargers. Daly reports that, "[A]s late as 1929, not a single student in the government secondary school was from Darfur, whereas there were 218 from Khartoum, 93 from Blue Nile, 41 from Kordofan, and even some foreigners" (Daly, 2007: 134-

135). The number of Darfuri who attended Gordon College is well documented with the first student from Darfur to enter the only institution of higher education in Sudan in 1934. As late as 1945, the policy of providing education above the level of elementary vernacular schools was considered by the central authorities to be excessive. All in all, Daly's account of the education policy in Darfur reveals the level of prejudice towards the rural Sudanese inherent in the colonial project. So, it should come as no surprise when Daly recounts that in 1935 the funding allocation to Darfur for the educational needs of a province with a population of 750,000 people was £E (Egyptian pounds) 1,200 per year, which was, "less than the salary of one senior government official" (Daly 2007: 136). Heather J. Sharkey argues that the colonial education policy led to a situation where the British favoured a small group of,

> self-defined Arabs at the expense of everyone else, including not only Arabic-speaking former slaves or their descendants, but also non-Arab Muslims and non-Muslims (such as the Dinka), the British cultivated a group of men who had the literacy and the political know-how to develop and articulate nationalist ideologies. Not surprisingly, these men defined a nation in their own social image, as an Arab Muslim community. (2007: 30)

Native administration, in effect, imitated the practice of segregation imposed by the *Closed Ordinances Act* but on a smaller scale. On a micro-level, indirect rule was designed to hermetically seal different tribal groups from each to facilitate greater control and decrease the cost of administration while on a provincial level *Closed Ordinances Act* aimed to maintain the stability of Darfur by preventing influences from the rest of the world from reaching the province. The British were never able to achieve this goal but what they did create was a detached population who were increasingly isolated from the formation of a national economic and political elite; an elite who would form the leadership of the anti-colonial struggle, and in time, come to rule the Sudan. This group would be limited to a small Khartoum-situated riverain elite who benefited most from the years of colonial government.

The lack of a wider nationalist movement encompassing the western and southern peripheries at the time of Sudanese independence was testimony to the success of the colonial policies of segregation and tribalization. By segregating and "tribalizing" rural Sudan the British were able to prevent the consolidation of a trans-regional alliance forming between the elites from the different regions of the country. Additionally, the policies blocked the emergence of a shared national identity across the Sudan. The vast majority of nationalist leaders were western-educated riverain Arabs. Darfur's elite were largely excluded from the preparations for independence. The absence of a western educated elite in Darfur precluded any meaningful Darfuri representation at the central level, which ensured the interests of the region were overlooked by the successive national governments after independence.

Darfur's political and economic leaders have continually struggled, and failed, to overcome the legacy of marginalization inherited from the colonial era. Darfur's separation from Sudanese political decision-making led to the formation of regional movements in the 1950s, 1960s, 1980s and then again in the lead-up to conflict in 2003. British designs to segregate Darfur from the rest of Sudan were clearly aimed to limit influences that would upset the stability of the native administration system in Darfur, but in doing so, the region suffered significant discrimination which has played out in the tensions between the centre and the western periphery since independence. Darfur also suffered from the racial prejudices of the British in Sudan which has left a mark on Sudanese politics in a similar way to tribal conflict and centre-periphery tensions.

Colonial Racial Ideology in the Sudan

Much has been written on the topic of race and racism as a feature of the colonial project and how colonial racism has transferred into the politics and culture of the modern era (Austen and Smith, 1969, Snowden, 1970, Adas, 1989, Wallerstein and Balibar, 1991, Said, 1992). This is not the place to rehearse the question of the causes of modern racism, except to say that the concept of race, as understood today, was a feature of the European conquest and control of the world. The history of European racism towards Black Africans is inextricably intertwined with the history of the Atlantic slave trade and slavery in the US, but it was in the 19th century that the racialist doctrines of an inherently inferior Black subject crystallized. The 19th century was a turning point in terms of the formulation of notions of racial differences, which Nancy Stepan (quoted in Wheeler, 2001: 33) notes when she argues that there was a "change from an emphasis on the fundamental physical and moral homogeneity of man, despite superficial differences, to an emphasis on the essential heterogeneity of mankind, despite superficial similarities" so that, "[B]y the middle of the 19th century, everyone [in Britain] was agreed, it seemed, that in essential ways the white race was superior to non-white races".

Despite the best efforts of scholars like Bernard Lewis (1992) to export the European version of racism towards Africans to the Arab world, there is a clear distinction between the Arab attitude towards Black Africans and the racialist framework that Europeans employed to understand their place in the world. As John O. Hunwick and Eve Powell have argued, "there is a lack of consistent literature that theorizes the inferiority of black people. Islam did not have its Gobineau" (2002: xx).

This different attitude is summed up neatly by Ali Mazrui who states that while "Arabs alerted the people of sub-Saharan Africa that they were black. Europe tried to convince Black people they were inferior" (2005: 70). In a number of ways, British colonialism was premised on a belief that Europeans were superior and that the inferiority of other races were measured against the characteristics inherently held by people from the British race. British colonialism, then, brought this specific form of racism to Sudan, introducing skin-colour as a marker of inherent racial qualities based on a Eurocentric typology, at times grounded in cultural superiority, at other times racially determined, but always the two were discursively indivisible.

It must be said that the relations between "Arabs" and "Africans" in the pre-colonial period were far from equal. The attitude of "Arabs" towards "Africans" was often based on prejudices, but not of skin-colour or on the grounds of racial difference. Black African slaves were commodities traded and owned by Muslims and despite some efforts to depict slavery in the Muslim world as being less brutalizing than European slavery, there was nothing benign about slavery in the Muslim world. However, there was a major difference in the practice of slavery in the East as opposed to the West. While many Black African slaves were captured, transported from Africa, and sold to Muslims in various regions of the Middle East, it was not a racial distinction which provided the legitimacy for the practice. African Muslims, regardless of the colour of skin, could not be enslaved. This is in stark contrast to the slave trade conducted by Europeans from Africa which revolved around the Black subject being naturally suited or ordained to slavery. In the East, physical differences were overlooked as the expansion of Islam into Africa suggests. The gradual conversion of African societies to Islam, and the Arabization of those societies, over a thousand year period is testament to the non-racialist foundations of Islam, and the potential for African societies to assimilate and adapt to cultural change ('Abdel Rahim, 1969: 6-7; Lapidus, 2002).

A similar process of Islamization and Arabization were responsible for the integration of the northern Sudan into the Muslim and Arab milieu. In the southern Sudan, conversion to Islam and appropriation of Arab culture were poorly received but some evidence of infiltration was apparent prior to the British conquest of the region. In Darfur, Islamization and Arabization had made considerable inroads by the 19th century with the royal family and the aristocracy Islamized and increasingly adopting Arabic culture and language. Amongst the Sultanate's subjects most were Islamized by the 19th century with Arabization spreading more gradually. Nonetheless, the Sultanate of Darfur had adopted a dual character, as on the one hand, it was Muslim and Arab, and on the other a sacral indigenous kingship. Muslim cultures, especially those in Africa, were comfortable with such dualism. It was the colonial powers with their sense of racial certainty that were unwilling to conceptualize such traditional dualisms that led to a profound change occurring within Sudan.

The Sudan, according to Janice Boddy (2008) was a special case in British colonial terms, because of the way that it was ruled as separate entities comprising one colony, and because of the competition between the British, the Egyptians, and the riverain Arabs for "civilizing" the "primitives" of the southern Sudan. Throughout Boddy's analysis the theme of the relationship between ethnography and the colonial system is most apparent. Boddy relates how the British administration encouraged ethnographic work that confirmed the cultural and genealogical differences between north and south, Arab and non-Arab in Sudan. In this way, the British were provided with what they argued were legitimate reasons for restricting the reach of both the Egyptian Arabs and riverain Arabs into the southern Sudan. H.A MacMichael, the author of a major ethnographic work himself, reveals the purpose of the "southern policy" was to provide an environment where "a series of self-contained racial units will be developed with structure and organisation based on the solid rock of indigenous traditions and beliefs… and in the process a solid barrier will be created against the insidious political intrigue which must in the ordinary course of events increasingly beset our path in the North" (Boddy, 2008).

Overall, one of the key themes that emerges from Boddy's work is that race was a means by which the Sudanese were systematically classified and ordered and that this process functioned as a corollary of colonial administration.

Heather J. Sharkey examines the contribution of British colonial policies to the racialization of Sudan by looking principally at education policy. Her starting point for the Sudan shares certain key similarities with those taken by Mazrui and Hunwick that pre-colonial Arab-African relations were not based on racial differences and that it was European colonialism that introduced racial theories into Sudan. Sharkey writes that "When British and Egyptian forces overthrew the Mahdist state in 1898 and established the Anglo-Egyptian Condominium, there was no such thing as self-identified 'Sudanese Arab' among the riverine region's Muslim elites. At that time, to be Arab was to be Muslim, to be Arab was to be free, and ideally, it was to claim an Arab pedigree" (2008: 29).

It is clear that race played a large part in the British imaginary of the Sudan and that under the British, Sudanese identities were increasingly defined by racial characteristics. Sudanese increasingly became either

Muslim and Arab or African non-Muslim during the colonial period. The Africanization and Arabization of Sudan was not a natural or organic historical process but one imposed by colonial power onto the various societies that were brought together by the British to form the Anglo-Egyptian Condominium of Sudan.

As mentioned, racial invention was a major part of the colonial project on a much wider scale. Colonial officials had no need to invent the racial profiles they used in the same way they invented tribal identities. By the time that the British were ruling over the Sudan, the invention of both the Arabs and Africans had solidified around a number of basic principles. Cromer's declaration on the Oriental mind is a concrete example of the British attitude to the Arabs over which they ruled: "The mind of the Oriental, on the other hand, like his picturesque streets, is eminently wanting in symmetry. His reasoning is of the most slipshod description... They are often incapable of drawing the most obvious conclusions from any simple premises" (Cromer quoted in Lockman, 2004: 93).

It might seem extraordinary, in the context of the orientalist attitude of Cromer and other British colonial administrators, to read the view of L.F Nalder Governor of the Northern Province from 1930-1936 when he describes Arab and British minds as inherently similar: "the culture of the north is one which is easily comprehensible to ourselves... Moreover the Arab mentality is not far removed from our own; we continually find things which surprise us, but seldom things which shock us. His ideas of right and wrong are broadly similar to our own" (Nalder quoted in Leinhardt, 1982: 25).

For Nalder, the African mentality was incomprehensible and their way of life baffling as he described Negroes as failing to grasp rational enquiry because "for him our laws of cause and effect have little existence or meaning" and furthermore because the Negro "lives in a tiny portion of a dimly realized earth capriciously interfered with by a spirit world that can be partially controlled by magic. In anything abnormal, anything strange, he realises the working of that world" (Nalder quoted in Leinhardt, 1982: 25).

However, Cromer's view is not incommensurate with that of Nalder's, but part of a wider British view of race and culture that was also shaped by the exigencies of colonial rule. In the case of the Sudan, the British rationalized colonialism as they did elsewhere, as a "civilizing mission". In this particular world-view, the ancient Egyptians, Babylonians and more recent Islamic empires had reached levels of civilization that had been at one time the most advanced in the world. The non-Arabs of Africa though were identified differently through a discourse traced back to the slave trade and which regarded Africans as a primitive race without history. For Francis Mading Deng, the British historian A.J. Arkell articulated the British view of Southern Sudan most aptly when he argued in 1955 that "what is known as the southern Sudan today has no history before 1821" (Arkell quoted in Deng, 1995: 185). The well accepted belief that Africans were a people without history was a recurrent element of the European discourse on Africa, continually repeated in the 19th century by the likes of Hegel and evident in the 20th century when Hugh Trevor-Roper Regius who became Professor of History at Oxford University in 1963 stated in his inauguration speech that Africa's history commenced with the European penetration. The African-Negro was subjected to a form of racism that not only implied an inferiority in relation to Europeans but also to Arabs and other Semites.

While Darfur is not in the southern Sudan and the vast majority of Darfur's inhabitants are Muslims the British demonstrated a certain ambiguity and uncertainty regarding the racial identity of the people they ruled over in Darfur. Over time, as Darfur was increasingly relegated to the periphery of the British interest in Sudan, the region came to be viewed with the same disregard as the southern Sudan. Despite, the gradual absorption and adaptation of Arabic culture and language by the inhabitants of Darfur's central regions, British ethnographic work framed Darfur's inhabitants as African Muslims. Darfur became a backwater of the Sudan, not because of its historical backwardness but due to the British neglect of the region which was justified on the basis of the racial character of the regions inhabitants. Margery Perham's views on Darfur illustrate this point.

Margery Perham visited Darfur in 1938 and her views on Darfur, while they are her views, do express something about the British colonial attitude. Perham taught and wrote extensively on Africa and occupied a senior position in the wider colonial infrastructure as Reader in Colonial Administration in Africa, at Nuffield College, Oxford University. In the first half of the 20th century Perham was a widely cited authority on Africa. Her teaching shaped the views of thousands of Britons who often were engaged by the colonial service in Africa. As a consequence, Perham's opinions are important for understanding the production and dissemination of knowledge relating to Africa and the attitude of the British during the late colonial period. Perham's diary entry following her visit to Darfur in 1938 illustrates only to well how far race shaped the way the British viewed the world during the late colonial period. She writes:

> We are in a great-wide basin, ringed by the peaks of pointed or domed rock. It all seems so un-African that I seemed so surprised, and perhaps, also disgusted, to meet negroes in this Paradise. For whatever I might feel about

> negroes, they are not romantic. Their appearance and their general servile character must always from this point of view, put them in a category apart from Red Indians, Polynesians, or Arabs (Perham quoted in Louis, 2006: 529).

Perham's rigid racism and racialism is evident from this quote. It is this rigid racialism that prevented the British from being able to comprehend the overlapping and essentially dualistic nature of identities in Darfur. Colonial administrators were determined to find an African or Arab identity in Darfur even when the distinction was completely arbitrary. But then, as Stuart Hall (1996) has explained only too well, racial distinctions are always arbitrary.

The "Africans" of Sudan were considered to be backward and their own "traditional' laws and customs, it was argued, would be more appropriate for them at the stage of development they had reached. Some minimal educational opportunities were provided for the chiefs and the sons of chiefs but otherwise it was felt education would "ruin" the indigenous populace. The introduction of racial justifications for the domination of one group over another has been profoundly debilitating for the Sudan. Racialism has been one of the primary fields on which political and economic struggles during the post-colonial period have occurred, and continue to occur. The north-south conflict and now the Darfur-Khartoum conflict have been infused with racial ideologies that can be traced back to the colonial perspective of races and racial hierarchies. The racial element in Sudan, as argued at the outset of this section, was a departure from the pre-colonial perceptions held in Darfur, which were shaped by Islam and the power of the Keira Sultanate.

Conclusion: the Colonial Legacy for Darfur and Sudan

Colonialism in Sudan followed a similar pattern of logic as in other colonial locations, even if the practices and outcomes differed. In Sudan, British colonial governments worked to secure the interests of empire and the colonial infrastructure. The major problem for the Sudan Political Service was how to maintain control over a vast land, where diverse people with few historical connections lived, and where the resources available to the state were meagre. The system of government that the colonizers adopted to manage the problems of distance, cultural diversity and a lack of manpower was based on tribalism, segregation and racism. Actually, the British did more than introduce new forces, but invented, manipulated and distorted the identities of the Sudanese to correlate with the British prejudices.

Darfur's conflict is inextricably connected to the tribal, regional and racial identities constructed by the colonial government during the first half of the 20th century. Therefore, it is with the historical processes engendered by colonialism that any examination of the causes of conflict in Darfur should commence and not with the often repeated simplistic notions that the conflict is the product of primordial tribal and racial hostilities. However, one must be careful not to fall into the trap of blaming British colonialism for the conflict in Darfur and equally careful not to ignore the colonial legacies for providing the context for understanding contemporary conflicts in Sudan.

Transnational Carceral Archipelagos

Lampedusa and Christmas Island

Joseph PUGLIESE

Associate Professor at Macquarie University, Sydney

> The present epoch will perhaps be above all the epoch of space. We are in the epoch of simultaneity; we are in the epoch of juxtaposition, the epoch of the near and the far, of the side-by-side, of the dispersed.
>
> (Foucault, 1986: 22)

Introduction

At a website dedicated to the reviews of tourists who have spent their vacation on the Sicilian island of Lampedusa in the Mediterranean sea, the island emerges, through these tourist reviews, as a "place of fable" with a "fairy tale sea", an "unforgettable postcard place" and an "island of dreams" (Tripadvisor, 2009). Reading these reviews, I experienced an acute sense of disorientation, as my vision of Lampedusa is of an island infamous for its overcrowded, traumatising immigration detention centre. It is a harrowing site that has witnessed the unsolicited arrival of hundreds of dead bodies of asylum seekers and refugees who, in their failed attempts to breach the borders of Fortress Europe in largely unseaworthy boats and rubber dinghies, have unceremoniously washed up on its shores. This essay marks an attempt to come to terms with these violently disjunctive accounts and experiences of (the same) space. The epigraph that mastheads this essay offers a point of entry into the complex and disorienting problematics of space inscribed by radically asymmetrical relations of power. The "epoch of simultaneity" articulates the possibility of temporally juxtaposing two absolutely dichotomous figures – the wealthy tourist from the Global North and the utterly disenfranchised refugee from the Global South – within the same geographical space (Lampedusa); this same space at once positions the "near and the far" "side-by-side" in order to bring into focus the fault line of the border that assembles and configures the "dispersed" in its matrix. As I discuss in some detail below, I will proceed to theorise the violently disjunctive experiences of space that inscribe a penal island/tourist resort such as Lampedusa by deploying Foucault's concept of *heterotopias*, a concept that enables the conceptualisation of absolute difference within the space of simultaneity.

My attempt to begin to unravel the complex texture of this heterotopic matrix is underpinned by a genealogical impetus. Precisely as Lampedusa is riven, in contemporary accounts, by the violently dichotomous fault line that divides along the axes of penal colony/tourist isle, Global South/Global North, another history altogether is insistently effaced and silenced: the history of Lampedusa, and its surrounding Mediterranean islands, as locus of a "carceral archipelago" established soon after Italian unification (Foucault, 1980: 68). This carceral archipelago was established by the northern dominated Italian government as the place to dispatch into penal exile the criminalised refuse of an insurgent, anti-unification South. Even as this genealogical history has been occluded in contemporary accounts of Lampedusa, I want to bring it into focus in order to mark the serial nature of penal camps as foundational to the process of modern nation building (Agamben, 1998; Perera, 2000).

In the course of this essay, I will proceed to examine the transnational dimensions of these carceral archipelagos, as paradoxical sites marked simultaneously as penal colonies and tourist destinations, in the context of past and contemporary Italian and Australian history. This transnational context, I argue, is inscribed by overlapping racialogical histories of whiteness, colonialism and empire, and attendant anxieties about securing a nation's borders. Lampedusa penal colony/holiday isle has, I argue, its equivalent in Australia's Christmas Island. Situating both these islands along the fault line of the border, I conclude by bringing into focus the border zones of the refugee dead that inscribe both these sites.

Colonial Genealogies of Penal Colonies and Carceral Archipelagos

Up until, and immediately after, the moment of unification, northern Italians viewed the South as a type of *terra incognita*. Knowledge of the South was largely gathered from the accounts and travelogues of European travellers who ventured into the nether regions of the peninsula and its islands and returned to tell their tales. Soon after the moment of unification, a collection of northern Italian politicians, bureaucrats and academics descended into the South in order to begin the process of integrating the region into the larger body of the Italian nation-state. This descent into the South was viewed in terms of a process of discovery and colonisation (Gramsci, 1975, 2005; Pugliese, 2007). A contemporary account published in a newspaper of Lombardy (one of the key loci of the unification movement) articulates the labour of discovery that lay ahead for a post-unified Italy:

> The first, the most important labour we need to accomplish as Italians is, and must be, the *discovery of Italy*. For most of us – and of course it is not our fault – Italy is a bit like Africa for geographers... we find ourselves in *terre ignote*... Therefore we say that we must rise to the task of *discovering* Italy (cited in Petraccone, 2000: 101).

Africa, in fact, became the governing metaphor through which Northerners made sense of the South, and the northern dominated nationalist government proceeded to deploy a colonial model that would establish "special careers with certain advantages like the House for Indian Affairs in England" (Petraccone, 2000: 101). The North's deployment of the governing metaphor of Africa instantiates the racialised marking of space that divides the Italian peninsula and its islands along a North/South, white/black axis. (Regardless of the fact that an island like Lampedusa is referred to, in contemporary histories, as *geologically* "part of Africa" but *culturally* Italian (Maps of the World.com, 2009), the South and its islands must be seen as also culturally North African (and Middle Eastern); as I have argued elsewhere, it is only through the enactment of strategies of white historicide that this cultural connection can be severed [Pugliese, 2007]).

In the course of the second-half of the 19th century, the governing metaphor of Africa, as a means whereby to make intelligible the differences between the South (represented as primitive, backward and barbaric) and the "civilised" North, was systematically fleshed out through the development of a series of racialogical theories propounded by a group of positivist anthropologists led by Cesare Lombroso. Renowned for developing the field of criminal anthropology and its theory of the "born criminal", Lombroso, and his positivist school, exerted a significant influence in the shaping of post-unification Italian legislative debates and policies on crime and law and order. Two key law and order policies directly influenced by Lombroso's theory of the "born criminal" were the *ammonizione* (warning) and the *domicilio coatto* (internal exile) (Gibson, 2002: 152). These two policing measures, informed by the positivist belief that "born criminals" could easily be detected both by their physical features and their behaviour, enabled a surveillance regime based on police discretionary powers that led to the imprisonment of "suspect" individuals in advance of having actually committed any offence.

The visual regimes that enabled the facile detection of criminally suspect figures were fundamentally informed by racialogical and racist assumptions. Listing, for example, the "abnormal" phenotypical features that distinguish a criminal from the "normal" citizen-subject, Lombroso writes that "These features recall the black American and Mongol races and, above all, prehistoric man much more than the white races" (Lombroso, 2006: 49). Situated in the context of these racist assumptions on the give-away "abnormal" "look" of the criminally suspect figure, Southern Italians were at the greatest risk of being innocently drawn into the net of the criminal justice system precisely because they were, in Lombroso's theories, already marked as biologically inferior types related to two other congenitally predisposed criminal races: the "Negroid" and "Mongoloid" (Pugliese, 2002: 157):

> Prognathism, thick and crisp hair, thin beards, dark skin, pointed skulls, oblique eyes, small craniums, overdeveloped jaws, receding foreheads, similarity between the sexes, muscular weakness – all these characteristics confirm the findings from autopsies to demonstrate that European criminals bear a strong racial resemblance to Australian [A]borigines and Mongols (Lombroso, 2006: 57).

The policy of *domicilio coatto* or internal exile, a policy established soon after Italian unification that largely targeted Southerners, led to the development of a series of penal colonies dispersed across the South's islands, including Lampedusa, Favignano, Lipari, Tremiti, Ustica, Pantelleria, Ventotene and Ponza (Gibson, 2002: 154). The policy of internal exile for Southerners was further buttressed by the passing of the Pica law (1863). The Pica law formally "legalised the state of exception" that saw the suspension, for Southerners, of all civil and legal rights, and the immediate execution of any Southerner found to have "insulted" the Italian flag (Martucci, 2007: 337, 324). Operating in tandem with the draconian anti-brigand Pica law, designed to repress and liquidate anti-unification Southern insurgents (Pugliese, 2008), the policy of internal exile effectively established what Foucault terms the "carceral archipelago": "the way in which a form of punitive system is physically dispersed yet at the same time covers the entirety of a society" (1980: 68). This carceral archipelago was dispersed across the Italian peninsula and its islands, and it included, as Luciano Salera documents, the

> 'holding' and 'concentration' camps that interned the dissolved *Regio Esercito Napolitano* [the Neapolitan kingdom's official army], the 'lager' Sabaudi of Genoa and Allessandria, and of Fenestrelle and San Maurizio; these all bring to conclusion the trilogy of massacres [of Southerners] perpetrated by the Piedmontese [who dominated the process of Italian unification] (2006: 32; see also Izzio, 1999).

The appalling conditions endured by the prisoners dispatched to the penal colonies of the South were documented in a number of contemporary sources. In her book on Lombroso's criminal anthropology, Mary Gibson cites the eye-witness accounts of an Englishwoman, Jesse White Mario, who travelled to the penal island of Favignana, off the Sicilian coast, and published a number of articles in the *Nuova Antologia* between 1896-97: "White Mario... noted the disproportionately high percentage of Sicilians in the penal colonies, showing the continuing prejudice and even outright persecution they endured at the hands of the central government. She concluded that the system of *domicilio coatto*, 'while evil in itself and harmful to all, is deadly for Sicilians'" (Gibson, 2002: 154). White Mario's concerns were corroborated by a report (1915) on the penal colony of Lampedusa from the Prefect of Agrigento to the headquarters of the Public Security police in Rome: "About 10 percent of the convicts had tuberculosis or syphilis, exacerbated by the filthy dormitories that needed disinfecting. He concluded pessimistically that the colony of Lampedusa was 'a preparatory school for the perpetuation and execution of future crimes since the infectious germ of criminality is disseminated with the exchange of ideas'" (Gibson, 2002: 154). The South's archipelago of penal colonies was viewed by the criminal anthropologists as supplying invaluable source material that could be used to validate their theories of congenital criminality. The penal colonies

> provided a literally captive population for experimentation. The *Bulletin* [a criminal anthropological journal] admitted as much, calling them 'precious human material' exhibiting 'a thoroughly antisocial essence'... Official reports from the school labelled... *coatti* [internal exiles] as 'the most fearful part of habitual criminality' (Gibson, 2002: 156).

Reading these historical reports that document the appalling conditions of such penal colonies as Lampedusa, I was struck by the manner in which they so clearly echo contemporary concerns about the humanitarian crisis unfolding in Lampedusa's immigration detention prison. An appeal, published on January 27, 2009, by multiple humanitarian organisations (including Amnesty International, Save the Children and *Medici Senza Frontiera*) has drawn attention to the 'alarming conditions' of the detainees in Lampedusa's immigration prison: "As has been said, the current situation sees over 1,800 people present in a centre whose maximum capacity is 804 places. It is glaringly obvious that this entails serious problems of overcrowding [and] worrying hygienic-sanitary conditions" (Progetto Melting Pot Europa, 2009: 1). UNHCR, the UN Refugee Agency, has also released a report documenting its concern over the appalling conditions that refugees and asylum seekers are being compelled to endure in Lampedusa's immigration prison (UNHCR, 2009).

Symptomatic of the process of maintaining "national security", these parallels between historically different situations and geopolitically disparate subjects (internal exiles and refugees) emerge as structural effects once situated within the biopolitical regime of the camp and the exclusionary racialogical theories constitutive of its operations. Drawing on Agamben's theorisation of the camp in the context of western modernity and colonial nation-building, Suvendrini Perera brings into focus the structurality that underpins these sites of historico-topological repetition: "Examining the juridical and political structure of the camp 'will lead us to regard the camp, not as an historical fact and anomaly belonging to the past... but in some way as the hidden matrix and nomos of the political space in which we are still living'" (2002: 1). The structural logic of the camp, as biopolitical matrix that pivots on maintaining "spaces of exception", enables the folding of one camp into yet another (Agamben, 2005). In the Australian context, Perera tracks the serial nature of these spatio-temporal folds: "The Australian camp is the site where the prisoner-of-war camp meets the long-term aims of colonial assimilation/annihilation in the forms of the outstation, the penal camp and the mission" (2002: 3). The carceral archipelago that was constitutive of Italian nation building has its Australian equivalent. Scattered across the offshore perimeters of the Australian continent are island gulags that have played a fundamental role in dispatching to internal exile the nation's designated criminal subjects. Australia's carceral gulags include, to name but a few, such sites as Rottnest Island, penal colony for Aboriginal political prisoners (Dovey, 1993) and, later, internment camp for Italian prisoners of war (Fontana, n.d.), Port Arthur and Wybelenna (Stratton, 2006), Palm Island (Rosser, 1978), Christmas Island and, in the wake of the so-called "Pacific Solution", the islands of Manus and Nauru (Perera, 2008). In his history of Palm Island, for example, Bill Rosser writes how, following the passing of the *Queensland Aboriginal Act* in 1884, the island became an Aboriginal penal settlement that was "considered to be an ideal place to confine 'problem cases' and 'uncontrolables,' mostly children" (Rosser, 1978: 2; see also Wadiwel, 2007).

In the Italian context, the penal colony of Lampedusa (established as a site to which to dispatch and exile insurgent Southerners from the body of the nation during the process of northern-controlled Italian nation building) has morphed into another form of island gulag with a contemporary immigration detention prison. On yet another topological level, in Lampedusa the "former NATO military base has been converted into a second refugee camp, but this one – known as an 'identification and deportation centre' – is patrolled by Italian military police and the *carabinieri*" (Totaro, 2009a: 11). Situated against this backdrop, in a striking moment of historical forgetting, Lampedusa's mayor laments that "We don't want to become the Alcatraz of the Mediterranean" (cited in Totaro, 2009: 11).

If the locus of the camp is demarcated as a "space of exception", then, as Perera argues, the inhabitant of this space is at once disenfranchised of her or his human rights and precluded from accessing due processes of law (2002: 2). The arbitrary and abusive exercise of discretionary police power, and the lack of recourse to due processes of law that characterised the regime of "internal exile" (*domicilio coatto*) for Southerners, could be characterised, as Gibson notes, "as preventive policing, since they purportedly defended society from potential crime" (2002: 157). Pursuing, in the Italian context, the serial effects symptomatic of the operational logic of the camp, the contemporary criminalisation of refugees and Roma peoples, as I discuss below, at once connects back to biopolitical regimes premised on "national security" concepts of preventative criminology and the attendant violation of human rights that this produces. A recent report by Amnesty International (that has been co-signed by a number of other prominent humanitarian organisations) documenting the "alarming conditions" in Lampedusa's refugee camps draws specific attention to the arbitrary and discretionary abuses of power unfolding in these prisons:

> In fact, it is well known that the legal framework that is in force and regulates rescue, assistance and early reception interventions for foreigners appears to have shortcomings and gaps, and is liable to be interpreted in different ways and be applied with discretion. In particular, the rights of foreign recipients of assistance measures in so-called early reception centres, which are still exclusively disciplined by Law 563/95 (referred to as the Apulia Law), are not defined. Several reports have highlighted how, often, foreigners are effectively held in the current early reception centres for considerably long periods that may vary from a few days to weeks or months, without the legal framework establishing the rights of foreigners who are there in a clear and binding manner, and without the situation of de facto limitation of personal freedom being subjected to any form of judicial scrutiny… this situation, which does not comply with Italian legislation in the field of measures limiting freedom… could be considered a violation of art. 5 point 1 of the European Convention on Human Rights (ECHR). (Progetto Melting Pot Europa, 2009)

Reflecting on the role of the camp in the context of the British colonisation of Australia, Perera describes "The imperial fantasy of a gulag continent where Britain's criminal classes could be permanently despatched"; in turn, this called for "the institution of a different kind of camp: an internment camp for the population they displaced" (2002: 3). This "imperial fantasy of a gulag continent" has its own Italian peninsular parallel. At the same moment (the opening decade of the 20th century) when the establishment of six new "colonies of confinement" (*colonie di relegazione*) in the southern regions of Calabria, Sardinia and Basilicata was announced, the Italian nation-state put forward a series of proposals for the deportation of "incorrigible criminals" to penal colonies in Africa (Gibson, 2002: 158). Working on criminal anthropological assumptions of the "born criminal", Prime Minister Luigi Luzzatti argued that "individuals unable to adapt to our civilization will adjust and feel at home in the semi-savage environment" (cited in Gibson, 2002: 158). As Gibson explains:

> Adopting the positivist equation of criminals and "barbarians", he suggested sending convicts and their families to colonize Africa, a plan that would provide maximum defense of Italian society at a minimum cost, since families would be self-supporting. Anticipating objections that such forced colonization would punish innocent family members, Luzzatti quoted the positivist maxim that "inborn criminality is almost never limited to one member of the family but rather is one of the types of anomalies that effects [*sic*] the entire stock". (Gibson, 2002: 158)

The racialogical theories of criminal anthropology are not something that can be dismissed as an example of anachronistic "junk" science that can be, in turn, merely relegated to the dustbin of the past. On the contrary, the positivist belief in the criminality of an "entire stock" is exactly what has informed the recent passing of legislation by the Italian parliament ordering the fingerprinting of all Roma people living in Italy, including children. "By arguing that born criminality was visible on the body", writes Gibson, "criminal anthropology encouraged the idea that police could correctly identify deviants even before they broke the law" (2002: 157). In contemporary Italy, police have been rounding up and fingerprinting *all* Roma women, men and children as a contemporary exercise in "preventive criminology", producing, in effect, an "ethnic register that would treat Roma children as if they were hardened criminals" (BBC, 2008). On a larger, transnational front, this criminalisation of suspect groups finds yet another equivalent in the deployment of the Eurodac project: "a European Union initiative to use biometrics (specif. fingerprinting) in controlling illegal immigration and border crossings by asylum seekers" (van der Ploeg, 1999: 295).

In the Italian context, the colonial dimensions of these racial-criminalogical regimes can be seen to emerge from the southern interior of the newly established Italian nation-state and exponentially fan southward in the process of Italy's imperial expansion into Africa. The impetus that led Lombroso to develop his theory of the born criminal can be tracked back to his stint, very early on in his career, working as a doctor in Calabria at the time of the civil wars that attended the struggle for Italian unification: "In an early positivist enterprise, he measured and studied 3,000 soldiers in Calabria in addition to his routine duties as physician" (Gibson, 2002: 19). Following his experiences in Calabria, Lombroso published *In Calabria* (1898 [first published 1862]), a text that set the course for later views of the South as an infernal place of congenital criminality, savagery, backwardness and filth. By the beginning of the 20th century, positivist anthropology had played its role in legitimating Italy's project of imperial expansion by transposing Lombroso's southern-inspired theories to Africa, offering, in the process, scientific legitimacy to colonial occupation through its racialogical theories. A so-called "identification" office was established in Libya in 1912, for "the purpose of submitting Libyans to the identification process": "These records of the physical attributes of its subjects would benefit Italy, whose colonial policy should take into account 'the biological and social laws of all societies, especially exotic ones'" (Gibson, 2002: 148). As Gibson concludes, Italy's positivist anthropologists were "delighted to have new data to bolster their racial arguments on the inferiority of Africans" (2002: 148).

The fascist regime's imperial ambitions further expanded the role of positivist anthropology as a way of legitimating its violent incursions into Africa by encouraging the transposition to the African colonies, for "racial purposes", of "anthropological-biographical" dossiers on "criminals": "Equipped with these dossiers, 'the Colonial Police Corps could improve their service, contributing efficaciously to the defense of the Italian race against possible infiltration from elements of different and inferior races'" (cited in Gibson, 2002: 151). This call to defend the "Italian race against possible infiltration from elements of different and inferior races" is, as I have discussed in detail elsewhere, fomenting the most virulent forms of white supremacism in the contemporary Italian context (Pugliese, 2008).

Crisis Heterotopias: Lampedusa/Christmas Island

In tracking the occluded racialogical history of the penal colony of Lampedusa, I have drawn attention to the structural parallels that hold between two very different nation-state formations, Italy and Australia, and their respective biopolitical deployments of the camp – as penal colony, space of exception and colonial instrumentality – functionally guaranteeing both the borders and security of the nation-state. These structural parallels must not be seen as things that are merely coincidental to both nations. Rather, I want to suggest that the history of the Italian nation-state's use of the camp is also constitutive of Australian history (and vice versa: Australia's immigration policy of mandatory detention has served, for example, as a model for contemporary Italian refugee policy, influencing the transformation of so-called "welcome" centres into immigration prisons [Totaro, 2009b: 7]), even as this other history is positioned as something entirely eccentric to Anglocentric accounts of Australian history.

The racialogical theories of Italian positivist anthropologists/criminologists, that were instrumental in establishing the theoretical models for Italy's internal and external colonial practices, were, as I have argued elsewhere, critical in influencing Australian immigration policy and its racially calibrated selection criteria, which pivoted on excluding non-white bodies from the corpus of the nation whilst also identifying prospective immigrants who were at least proximally white and thus amenable to being assimilated into the white corpus of the nation (Pugliese, 2002). Furthermore, the Italian subjects who entered Australia did not abruptly leave their own embodied histories of previous regimes of racism and colonialism on their departure from Italy; on the contrary, these embodied histories effectively enabled the reproduction, for example, of the very racist hierarchies that had been inculcated back in Italy and the reinscription of Italian colonial practices on Australian soil. I evidence the functional coextensiveness of Italian colonial practices with their Anglo counterparts by mentioning two colonies established by Italian migrants in the late 19th century: Maria Island or Toarra-Marra-Moha to use its original Aboriginal name, off the coast of Tasmania, founded in 1884 by a group of Italian immigrants on the former British penal colony; the establishment of this new colony on Maria Island ensured the second-wave dispossession of the Tyreddeme people of the Oyster Bay nation; and New Italy, founded in 1885, on the north coast of New South Wales; once again, this Italian colony enabled a second-wave dispossession of the country's traditional owners: the people of the Bandjalung nation.[1] In *My Bundjalung People*, Ruby Langford Ginibi documents the successive waves of colonial invasion that ensured the dispossession of her people and the instrumental role of camps and missions in consolidating this process (1994: 87).

Both these colonial projects, the Italian colonies of New Italy and Maria Island, furthered the ongoing colonial expropriation and usurpation of country from the traditional Aboriginal owners. To a degree, these other histories of non-Anglo colonialism, which demonstrate the manner in which non-Anglo immigrant colonial practices and values were effectively transplanted and grafted onto existing colonial formations, have been largely ignored in both Anglo and non-Anglo Austra-

1 The brief discussion staged here of the colonies of Maria Island and New Italy are part of a larger research project in which I am examining the systemic relations between Italian and Australian colonialism.

lian histories. What I am attempting to bring into focus here is the transnational nature of biopolitical and colonial regimes, and the manner in which the transnational infrastructure of biopolitical and colonial regimes ensures the reproduction, across different sites and bodies, of violent relations of power. The grafting of prior, transnational colonial practices onto existing Anglo colonial formations ensured both the reproduction and the consolidation of the dispossession and disenfranchisement of Australia's Indigenous peoples.

Transposing this analysis from the past to the present, I want to focus on two locations that crystallise these transnational relations of biopolitical power: Lampedusa, off the Sicilian coast, and Christmas Island, off the Western Australian coast. Both islands, situated in completely different geographical zones, are marked by similarly complex biopolitical topologies and economies: they are at once contemporary penal colonies established for the detention and processing of refugees and asylum seekers *and* tourist destinations and holiday isles. It is this apparently irreconcilable contradiction that I want to analyse.

I opened this essay by citing some of the traveller reviews of tourists who have spent their vacation on the Sicilian island of Lampedusa in the Mediterranean sea. In these reviews, Lampedusa emerges as a "place of fable" with a "fairy tale sea", an "unforgettable postcard place" and an "island of dreams" (Tripadvisor, 2009). Similarly, Christmas Island also has its own dedicated tourist websites (which paint the picture of an "Island paradise": "Christmas Island is world-class. 250 square miles of coral sand, iridescent lagoons and coconut palms" [Trussel, 1988: 1]). At the web page dedicated to traveller reviews of Christmas Island, tourists wax lyrical about the island offering "spectacular sunrises", "a quiet retreat set on fabulous gardens", "the place is ideal for 'getting away'" (Tripadivsor, 2004). As I mentioned above, reading similar reviews of Lampedusa, I experienced a sense of disorientation, as my vision of Lampedusa is of an island infamous for its overcrowded, unsanitary immigration detention centre and as a harrowing site that has witnessed the shipwrecked bodies of hundreds of asylum seekers and refugees washed up on its shores.

Similarly, Christmas Island's history is harrowed by the presence of its own $400 million Immigration Reception and Processing Centre, which incarcerates not only men and women but also infants and children: "A pregnant woman is among 454 people detained on Christmas Island and medical professionals warn of potential health repercussions for the mother and child. The detention centre on the island is now home to at least 45 children aged one to 17" (Narushima, 2009: 5). The ongoing imprisonment of refugee children in Australia's immigration prisons is in direct violation of the United Nations Convention on the Rights of the Child, to which Australia is a signatory. Christmas Island, furthermore, is a site that has been unforgettably inscribed in the *Tampa* controversy, which saw the turning away of 438 asylum seekers rescued by the Norwegian ship of that name, and the detention of the asylum seekers after the so-called "children overboard" scandal, in which the government falsely claimed that the asylum seekers had wilfully thrown their children into the sea from their boat (when in fact it was sinking) in order to compel the navy to rescue them (Perera, 2000a, 2006). In his ironically titled essay, "Dying to Come to Australia: Asylum Seekers, Tourists and Death", Jon Stratton meticulously documents the perverse and violently contradictory relations that at once dis/connect asylum seekers and tourists within political economies founded on "white Australia's anxieties about people coming to the country" (2007: 180).

By invoking Foucault's "epoch of simultaneity", as cited in my epigraph, I want to stage the temporal juxtaposition of two absolutely dichotomous figures – the wealthy tourist from the Global North and the utterly disenfranchised refugee from the Global South – within the same geographical space (Lampedusa and Christmas Island). This same space at once positions the "near and the far" "side-by-side" in order to bring into focus the fault line of the border that assembles and configures the "dispersed" in its matrix. Foucault's conceptualisation of an "epoch of simultaneity" enables the possibility of thinking, at one and the same time, of the aporetic coexistence of the penal/holiday island and the refugee/tourist, precisely as it proceeds to articulate "an ensemble of relations that makes them appear as juxtaposed, set off against one another, implicated by each other – that makes them appear, in short, as a sort of configuration" (Foucault, 1986: 22). Foucault proceeds to name these complex configurations of space/time "heterotopias". Heterotopias, he explains, stand in contradistinction to utopias (entirely mythic and unreal places): they are other places that "exist in reality" even as they exert "a sort of simultaneously mythic and real contestation of the space in which we live" (1986: 24). At once grounded in reality yet invisible to some, at once marked by lived experience yet unreal because they are *other* spaces that remain unintelligible to the dominant order that governs a space, heterotopias configure complex and contradictory dimensions of space-time.

In his catalogue of the various forms that heterotopias can assume, Foucault designates two heterotopic formations that I want particularly to examine: crisis heterotopias and heterotopias of deviation. "In the so-called primitive societies", writes Foucault, "there is a certain form of heterotopia that I would call crisis heterotopia, i.e., there are privileged or sacred or forbidden places, reserved for individuals who are, in relation to society and to the human environment in which they live, in a state of crisis: adolescents, menstruating women, pregnant women, the elderly, etc." (1986: 240). Regardless of the vitiating qualifier, "so-called", the gendered colonial discourse of primitivism persists, in Foucault's account, in framing the subjects who inhabit crisis heterotopias. The other heterotopic formation that Foucault identifies is what he calls "heterotopias of deviation: those in which individuals whose behavior is deviant in relation to the required mean or norm are placed. Cases of this are rest homes and psychiatric hospitals, and of course prisons" (1986: 25).

I want to transpose Foucault's concept of crisis heterotopias, as spaces that simultaneously fold into heterotopias of deviation, and resignify them – so that they delineate, specifically and insistently, the spaces inhabited by those marked as deviations from the norm: refugees, asylum seekers and the "undocumented" from the Global South that are branded as "unlawful", "illegal" and "unauthorised" because they have failed to go through the "proper" institutional channels in their seeking of asylum. At yet another level, I want to resignify crisis heterotopias by relating Foucault's notion that they designate spaces of the "sacred" to Giorgio Agamben's (1998) theory of the camp as the space within which the sacral figure of *homo sacer* emerges, lives and dies. In Agambenian terms, *homo sacer* is the embodiment of 'bare life": a figure utterly disenfranchised of human rights and due processes of law; *homo sacer* is a sacrificial figure that can be killed with absolute impunity. The sites of Lampedusa and Christmas Island must be seen, furthermore, as crisis heterotopias riven by the violent logic of "crisis capitalism". Drawing on Naomi Klein's theory of "disaster capitalism", Stratton has mapped the manner in which "neoliberalism has been facilitated by the opportunistic use of crises to enable the imposition of its economic policies" (2009; see also Perera, 2009). In "Uncertain Lives: Migration, the Border and Neoliberalism in Australia", Stratton has focused on Australia's refugee crisis and its relation to the manipulation of "Australia's population anxieties and fears around race and immigration" in order to further, in quite spectacular ways, nefarious political ends (Stratton, 2009: 679).

In the context of two very different nation-states, Italy and Australia, crisis capitalism, and its neoliberal political apparatus, has designated two islands, Lampedusa and Christmas Island, as fault line spaces marking the borders of Fortress Europe and Fortress Australia. Precisely as Lampedusa delineates the fault line between Europe and Africa, Christmas Island marks the fault line between Australia and Asia. And, as with Lampedusa, Christmas Island is inscribed with its own complex history of European colonialism and annexation (Perera, 2002a: 25-6). As Hamish McDonald writes, Christmas Island "was once attached to

Singapore, but ahead of the colony getting independence, Australia convinced the British to transfer sovereignty to Canberra in 1957 to keep the island's phosphate reserves and strategically useful airfield in safe hands" (2009). Both Lampedusa and Christmas Island are geographically and culturally connected to the very continents, Africa and Asia respectively, that they have now been mobilised to thwart and exclude. Situated within the transnational relations of colonial expansion that inscribe the histories of both Italy and Australia, McDonald's ironic observation on the contemporary results of past colonial ambitions applies to both nation-states: "A preoccupation with maritime territorial aggrandisement... moved the beachhead for potential refugees a lot closer to the South-East Asian [and African] jumping-off points" (2009).

Both Italian and Australian governments have capitalised on the fomenting of fear and anxiety with regard to the landfall of refugees and asylum seekers on these islands. These two penal islands bring into sharp focus the geographic outer limits of the nation even as they enunciate, in Stratton's words, "the distinction between the order of the practice of the nation-state and the disorder of the state of exception that lies beyond it – and yet is already implicit within it" (2009: 682). As instantiations of crisis heterotopias, founded on economies of crisis capitalism, these two islands materialise the state of exception in the topological form of the immigration detention prison. These island gulags arrest the deviant life forms of the Global South that wash up on their respective shores and imprison them within the crisis heterotopia of the immigration detention jail.

On another level, precisely because these two islands are productive of heterotopic spaces that enable the "juxtaposing in a single real place several spaces, several sites that are in themselves incompatible" (Foucault, 1996: 25), both Lampedusa and Christmas Island continue to signify as paradisaical tourist destinations, with luxury resorts and, in the case of Christmas Island, even a casino. In the schema of heterotopic space, the absolutely other space, the penal colony, becomes invisibilised and unintelligible within the enframing discourse of western tourism: spectacular sunsets, sparkling waters and serviced luxury – all effectively work to disappear the squalor and suffering of the immigration prison just over the horizon. Foucault identifies the temporal dimensions of these tourist heterotopias as linked "to time in its most fleeting, transitory [mode]... to time in the mode of the festival" (1986: 26). He offers the example within which "a new kind of temporal heterotopia has been invented: vacation villages" (1986: 26). Vacation villages unfold the temporal mode of the festival, and the temporary inhabitation of a completely other life (Foucault offers the example of the "Polynesian village") that allows the western tourist to shed the burdens and care of the industrial city and to go "native".

This tourist experience of fleeting, festival time must be juxtaposed with the no-time of indefinite detention that hollows out the daily existence of the refugees imprisoned within the immigration penal colonies. Heterotopias emerge, then, as sites marked by heterochronic dimensions that fold dichotomous experiences of space into incompatible dimensions of lived time. On Lampedusa and Christmas Island, these violently disjunctive heterochronies (interminably suspended carceral time versus festival or vacation time) unfold across violently disjunctive heterotopias, the space of the prison/resort. These two incompatible orders of space-time fold silently, invisibly, one into the other yet never breach their respective borders. In all the traveller reviews documented on the tourist websites of Lampedusa or Christmas Island, the presence of the immigration detention prisons remains unremarked. Their unspeakable status within these tourist narratives only serves to bring into haunting focus the weight and solidity of this extraordinary absence/silence. Nothing more graphically exemplifies the manner in which incompatible heterotopias and heterochronies co-exist within the one locus, yet remain seemingly invisible to each other, than the images of the washed-up corpses of refugees on the beaches of Europe in the line of sight of tourists continuing to enjoy their vacation, absolutely indifferent to the dead bodies on the beach. Peter Hulme describes one such image:

> One particular photograph shows a young couple, presumably Spanish, almost certainly European, in swimming costumes, sitting on the beach at Zahara under a parasol. A few yards away lies a third figure, the body of a would-be refugee or migrant, drowned on the attempted crossing of the straits of Gibraltar and washed up onto the beach. The couple gaze with apparent indifference in the direction of the dead man, their own bodies betraying no evident discomfort or anxiety. If the corpse were removed from the picture, it would just look as if they were enjoying a pleasant day at the seaside. (2005: 43-44)

The bodies on these beaches are not located on just any stretch of the European coastline. On the contrary, as Hulme underscores, they are located on some of "the contemporary globe's most significant fault-lines" (Hulme, 2005: 58). The beach at Zahara straddles the fault line of the Strait of Gibraltar, site of countless attempts by the subjects of the Global South to cross into Europe. The disturbing photograph that Hulme discusses has its chilling Italian equivalent. A photograph published in July 2008 shows another European holiday couple enjoying their day at a southern Italian beach (Torregaveta), totally indifferent to the presence of the drowned bodies of two young Roma girls, Violetta, 12, and Cristina, 13. One eyewitness describes the scene:

> Their bodies were left on the beach for an hour before being collected, just covered by a beach towel while people just got back to sunbathing and playing football. It was very surreal. There was this picture of a typical Italian beach with families enjoying the sun and then just metres away were the bodies of these two children. People were completely indifferent about what had happened (cited in the *Herald Sun*, 2008).

These photographs of two southern European beaches, that simultaneously represent holiday makers and dead subaltern subjects, dramatise the strange logic of what Perera terms the "borderscape": "the multilayered spaces of this unstable border zone", "with its crossing and recrossing of territorial and temporal boundaries" (2008: 206). As complex borderscape formations, they exemplify the heterochronotopes that govern the lives and deaths of their different subjects, precisely as they bring into focus the fraught and contradictory relations between bodies and borders. The dead and alive bodies that inhabit these photographs establish points of connection between incompatible yet simultaneous heterochronotopes: the heterotopias of the beach as holiday space *and* of the beach cemetery; the heterochronic modalities of the dead time of holiday languor *and* the dead time of rigor mortis, necrosis and decomposition. Within these space-time continuums that overlap but remain heterogeneous to each other, different ontologies of living and dying are experienced by their respective subjects. These conflicting and incompatible ontologies of living and dying are, as I have discussed elsewhere, reproduced through a biopolitical series of border chronotopes – the airport, the port, the railway station, the immigration and customs checkpoint – and technologies of trade and transport – the jet, ship, train, truck, shipping container (Pugliese, 2009: 160-64). These border chronotopes dramatise the two-tier ontologies that organise bodies along the Global North/South axis: simultaneous in their spatio-temporal unfolding, they produce dispositions of bodies and subjects that remain proximate yet invisible to each other, even as they generate radically different modalities of living and dying. Within these disjunctive heterochronotopes, subaltern subjects (the refugee, the undocumented, the *clandestino*) live and die the ontotautology of their clandestine status, literally remaining invisible to First World subjects *despite being directly in their line of sight.*

Marked by their pariah status, these two dead Roma girls signify nothing more that mere beach detritus. Inscribed by white supremacist racial hierarchies, they are assigned their rightful place amidst the beach litter, as they cannot be seen as possessing any proper claim to the category of the "human" subject. The assignation of such outlaw bodies as nothing more than environmental pollution contaminating the ecology of the tourist beach has been spelled out by the mayor of Lampedusa, Bruno Siragusa, angry at the unsolicited arrival of dead refugee bodies on the holiday beaches of his coast: "he has complained about the number of bodies being washed ashore. He says Lampedusa needs additional boats from Sicily to help clean up the waters, because the corpses are spoiling the tourist trade" (Waugh, 2005: 11).

Heterotopias, Foucault argues, are neither open-ended nor freely accessible. Rather, "Heterotopias always presuppose a system of opening and closing that both isolates them and makes them penetrable. In general, the heterotopic site is not freely accessible like a public place. Either the entry is compulsory, as in the case of entering a barracks or a prison, or else the individual has to submit to rites and purifications" (1986: 26). The crisis heterotopia of the immigration prison is violently marked by the compulsory entry of its refugee and asylum seeker population, escorted by either military or police figures. The tourist heterotopia is, on the contrary, distinguished by the self-selecting principles of capital and wealth, that effectively determine entry into the resort in the first instance, and that consequently ensure that the privileged visitor can proceed to undergo the rites of purification enabled by such facilities as the spa, the swimming pool and the therapeutic massage.

Boat Heterotopias and Littoral Zones of the Dead

> The boat is a floating piece of space, a place without a place, that exists by itself, that is closed in on itself and at the same time is given over to the infinity of the sea.
>
> (Foucault, 1986: 27)

In the closing lines of his essay on heterotopias, Foucault declares that "The ship is the heterotopia *par excellence*" (1986: 27). Inserting this assertion within an uncritically colonialist and gendered discourse, he celebrates this maritime technology in terms of a roving and unfettered entity

> that goes from port to port, from tack to tack, from brothel to brothel, it goes as far as the colonies in search of the most precious treasures they conceal in their gardens, you will understand why the boat has not only been for our civilization, from the 16th century until the present, the great instrument of economic development… but has been simultaneously the greatest reserve of the imagination (1986: 27).

Situated within the concerns of this essay on "boat people" or asylum seekers and the penal colonies in which they are so often incarcerated on landfall, this celebratory colonialist account of the boat heterotopia fails to resonate. Rather, Foucault's other conceptualisation of the boat heterotopia – as "a place without a place… given over to the infinity of the sea"– evokes, in both Australian and Italian contexts, the picture of shipwrecked boats and dinghies and the tens of thousands of lives of refugees, asylum seekers and immigrants lost in the no-place of the infinity of the sea (see United for Intercultural Action's List of 11,105 documented deaths through Fortress Europe).

Suvendrini Perera has documented the manner in which,

> Since 2001, Australia's surrounding waters have seemed awash with bodies: the phantom bodies of the children who were not thrown into the water (as suggested by a false government report issued to discredit asylum seekers); the unknown bodies, mostly children, lost when SIEV X, the unknown boat in a sequence of such vessels, went down with 353 asylum seekers on board (2006: 638).

Perera proceeds to unfold the tragic story of the SIEV X by bringing into focus the series of government lies and occlusions that continue to provoke "serious questions about the unknown role played by the covert disruption program funded by the Australian government in Indonesia [from which the SIEV X set off for its journey to Australia]" (2006: 646). In the face of the mire of government denials and cover-ups that haunt this maritime disaster, Perera presents the harrowing testimonials of some of the survivors of this shipwreck as "bodies of evidence" that speak otherwise of this disaster, even as they compel the assumption of responsibility for this loss of life.

In the course of this essay, I have been tracking what appears to be a series of parallels – Lampedusa/Christmas Island, penal colonies/holiday isles, the corpses on the beaches of Zahara/Torregaveta and so on – arguing, in the process, that these parallels must be seen as structural and not coincidental within transnational biopolitical economies. Following in the wake of Australia's SIEV X disaster at sea, I want to conclude this essay by staging one more parallel. Chronologically prior to the SIEV X disaster, on the night of the 25 December 1996, a rusting boat carrying 450 South Asian asylum seekers collided with a ferry in the Strait of Sicily. Described as "the worst shipping disaster in the Mediterranean since the Second World War" (*The Independent*, 2001: 1), 283 asylum seekers died. For four and half years, the survivors "struggled to convince the Italian authorities that 238 of their companions had been killed in a collision off the coast of Sicily" (*The Independent*, 2001: 1), despite the fact that Sicilian fisherman, soon after the shipwreck, began to haul in their fishing nets whole corpses and body parts of the victims of this maritime disaster:

> A fisherman in the village of Portopalo has told the Rome daily newspaper *la Repubblica* that he pulled a corpse out of his nets in early January 1997. The man said he had felt compassion but then remembered another fisherman who had once lost days of work after handing in a body. "I picked it up

and heard a thump, the head had fallen off the neck", he recalled. "I closed my eyes and heaved it overboard, then I gathered up the head and threw it back too". A local politician, priest and historian confirmed that many fishermen had had similar experiences (*The Independent*, 2001: 2).

In the course of the fisherman's work, the drowned corpse of a refugee hauled up in the net amounts to little more than human cargo that temporarily supplements his catch of the day, only to be unceremoniously discarded as stock not worth the cost of bureaucratic red tape and lost wages. As such, these bodies are stripped of their human status (an anonymous torso, a faceless head, fragmented body-bits disconnected from the legal category of rights-bearing person) and thus become utterly coextensive with their marine environment, dissolving into what I have elsewhere called "bodies of water" (Pugliese, 2006: 18).

Despite the testimonies of the survivors of this shipwreck, and the "bodies of evidence" brought to light by the fishermen's nets, the Italian authorities continued cynically to describe this disaster as a "phantom shipwreck" (*naufragio phantasma*) and a "presumed shipwreck" (*presunto naufragio*) until a fisherman, in June 2001, years after the event, hauled up in his fishing net the laminated identity card of the seventeen-year old Tamil victim, Medialagan Ganeshu, whose brother had also drowned (*La Repubblica*, 2002). In the eyes of the Italian authorities, the oral testimonies of the survivors had failed to carry any evidentiary status. It was only the existence of Ganeshu's identity card that led to the commencement of an official investigation into this tragedy. These bodies of water literally signified nothing: invisibilised, transparent, they were mere phantoms that could neither represent themselves nor be represented. Their death was not death; rather, it was a presumed death, a phantom death, or only the hysterical and untrustworthy babble of would-be immigrants who had tried to breach the law of the border. It was only the survival of Ganeshu's shard of official bureaucratic identity that could provide the proof of his death and the confirmation of the reality of the larger suppressed and effaced disaster.

From the depths of this watery necropolis of dismembered bodies, marine necrosphere of the refugee dead, a mere trace of Medialangan Ganeshu is brought to the surface. Caught in a net of glistening fish, it emerges as a proxy that can speak the unspeakable and name the unnameable: here is the bureaucratic simulacrum of the corpus of Medialangan Ganeshu; it is more real than his body of water. His photo, birth date, phenotypical attributes and signature carry more weight than the necrogenic flesh that supplemented the fisherman's harvest. This calling card of the dead will attest to what you had dispatched to the unknowledge of the abyss: that 238 refugees died in the Strait of Sicily at the time that you sat down to enjoy your Christmas dinner.

These stories of refugee maritime disasters and of innumerable deaths at the border reconfigure Foucault's celebratory boat heterotopias sailing on their infinite seas into crisis heterotopias of loss, death and mourning. These refugee stories tell of fatal border zones – the Strait of Gibraltar, the Strait of Sicily, the Southern Italian coast, the Western Australian coast, Ashmore Reef – precisely as they document the littoral repositories of the countless refugee dead, the unmarked grave sites in which the drowned victims become mere bodies of water.

Whose Whiteness?

Cultural Dis-Locations Between Italy and Australia

Gaia GIULIANI

Postdoctoral fellow, University of Bologna and
Associate visiting scholar, University of Technology, Sydney

Introduction

At the turn of the 20th century in Australia, internationally accepted anti-black stereotypes pervading Australian racial thinking and race-relations, resulted as revitalized and even transfigured by both social Darwinism and Eugenics. In their new formulations, either in terms of environmentalist theories of human evolution, or those of biological heredity, they conveyed a restricted conception of whiteness that systematized the exclusion from the *body politic* of a number of subjects: first of all Aborigines and, amongst incoming migrants and indentured labourers, Asians (Chinese, in particular), and Mediterraneans. In the project of a White Australia, Aborigines were excluded as impeding a full physical and symbolical appropriation of the continent as a *white* space; Chinese migrants as "invading the country" and weakening white numerical supremacy; and *not-so-white* Southern European and Mediterranean people of more recent immigration, as their dubious whiteness could undermine the supposed *purity* of British and North European descendants.

This essay explores some aspects of the idea of whiteness as they have emerged, within the Australian context, from the international debate on race of the end of 19th century and up to the 1930s. It will focus on the Australian inheritance and re-articulation of the oldest and most deep-rooted prejudices against blacks and not-white races.[1] In

[1] On the genealogy of those stereotypes see: Landucci (1972) on Enlightenment and colonial Others; Stannard (1992) regarding the New World; Jordan (1977) in reference to the British empire's anti-black stereotyping; Fredrickson (1971) on U.S. anti-black stereotypes; Stepan (1982) and Anderson (2007) on the British empire's racial stereotypes against "inferior races" and Aborigines.

particular, this essay will investigate the influence exerted on the Australian intellectual and political establishment by the theories of the inferiority of Southern Italians, and brown and black races as articulated by the Italian criminologist Cesare Lombroso (1835-1909) in the 1870s-1890s. This investigation will help reveal the similarities between the discursive and material aims and strategies related to Italian "internal colonialism" and Australian settler colonialism.

Additionally, the essay will contextualise these influences within the frame of the scientific approaches and outcomes of medical field-studies undertaken by one of the most authoritative Australian doctors of the 1930s, Raphael Cilento (1893-1985), whose research in the Tropics made a crucial contribution to the Australian debate on hygiene and racial purity. Contextualisation will allow me to demonstrate Italian-Australian differences and similarities within scientific as well as ideological approaches to race.

In order to grasp these peculiarities, and connect them to both a general theory of racism and the historical frame they belong to, I draw on Colette Guillaumin's influential categorization in *L'idéologie raciste* (1972) of two distinct genealogies of the racist phenomenon, originating from different racial identity-politics: *auto-referential* and *hetero-referential* racialisation. These categories, as Guillaumin stressed, never appear in their "pure" form, but are always intertwined, permeated with particular exigencies and aims, and with the cultural and social structures in which they are entrenched. A survey of early federal Australia's *auto-* and *hetero-* racial identity-politics is crucial to understanding the set of factors behind Australia's cultural and social permeability to science-based racial stereotypes, and in particular the role played by historical contingencies, national/ist imaginaries and State's policies.

Guillaumin's categories will help us to understand why particular (international) racial theories had been preferred over others, and to trace the trajectories of those international theories themselves. They will help us to appreciate the theoretical interchange between Italian *fin-de siècle* and Fascist *raciologies* and the Australian intertwining between environmentalist theories and eugenics, and consequently to stress the transcontinental nature of late 19th and early 20th century racialised "infoscapes", "ideoscapes", "(soma)technoscapes" and "ethnoscapes" (Appadurai, 1996: 33). Finally, in stressing cultural dislocations and disjunctural processes between Europe and Australia, Guillaumin's categories will show how European, and specifically Italian, *raciologies* (Gilroy 2000), transfigured and situated in the Australian cultural and political context of the *Immigration Restriction Act* (or White Australia Policy 1901), has been functional to a conception of the Australian political subject as "purely white".

Race Thinking and the Sovereign Subject in Early Federal Australia

In early 20th century Australia, the symbolic and material construction of a national sovereign subject (the Australian People) was in its early stages and needed a refined definition of the national Self – "God give us men – sagacious men – to make this latest born of nations truly great" (Alomes, Jones, 1991: 197, and broadly 182-201). The sovereign subject needed to be discursively and materially constructed as both a constitutive foundation and a constituting engine of the New Nation. This reformulation took the form of a self-attribution of cultural and bodily qualities, which were to be the measure of the political subject's cultural, political, physical, genetic and phenotypic uniqueness and incommensurability with other social (racialised) groups inhabiting and coming to Australia.

In this context, the late 19th and early 20th century re-elaboration of contradictory sets of trans-national epistemic systems,[2] from broad environmentalism to social Darwinism and eugenics, was crucial. This reconstructed and "situated" reformulation allowed the State's and society's agents to fulfil what elsewhere has been called *White Australia's utopia* or, in Edwin James Brady's words, "Australia unlimited" (Melbourne, 1918). They allowed those agents to define themselves through processes of "auto-referential" and "hetero-referential" racialisation (Guillaumin, 1972: 6) based on scientific evidence.

In Australia, in fact, the self-assumption of *humanness* (very similarly to the process of self-Arianization described by Leon Poliakov in 1971) that excludes all the other races from being considered *human* (auto-referential racialisation) appeared intimately intertwined with its opposing process, that of the assumption of the others' embodiment of diversity (hetero-referential racialisation). The latter process could be considered very similar to what Gayatri C. Spivak called "Othering" (Spivak, 1985: 132) although strictly founded on a scientifically based idea of race. This complicated intertwining between auto- and hetero-racial definitions – which we could define, in Ruth Frankenberg's words, as the effect of "the simultaneous operation of race and racism" (Frankenberg, 1997: 9) – seemed to respond to the need of hypostatising the exclusive British or Anglo-Celtic "possession" of whiteness against not-so-white migrants (see Lipsitz, 1998: vii-viii; Anderson, 2002: 158-161); the exclusive possession of Australian land by white colonizers (through the deployment of the *Terra Nullius* principle and the doctrine

---

2 For a distinction between theoretical (or ideological) racialism/racism and popular racism and their continuous interplay I refer to Wieviorka (1991).

of the *non-humanity* of Aboriginal Australians); and the white norm as the unique norm ruling over Australian social relations (Wolfe, 1999: 28-29). As such, the intermixing of auto- and hetero-referential racial identity politics seemed to respond to the need of imposing a viable scheme of inclusion/exclusion and *divide et impera* devices over a number of black, not-white and differently white subjects.

Specifically, in the case of migrants' racial differences the "hetero-referential" racialisation served as a legitimating discourse for rejecting migrants at the frontier, discouraging foreign governments in allowing emigration to Australia, and reproducing a racialised and often segregated labour force inside its borders. All this by virtue of a complicated interplay of capitalism, nationalism and "white fantasies" – as a number of scholars have suggested in a national and transnational perspective (see Giuliani, 2010).

Within this framework, the large influence that Lombrosian theories had outside Italy and particularly in Australia becomes clearer. These theories influenced the migration policies of a number of different nations, as Joseph Pugliese outlined in his essay "Race as Category Crisis" (2002: 157) as well as on the so-called "colonial" scientific and medical disciplines predicating social Darwinism. One of the main reasons of their influence lay in their methodology. Skull and body measurements (especially craniometry) as theorized and practiced in Lombroso's research on criminal attitudes and the a-sociality of women (1892), Southerners (1876; 1898), blacks (1871) and anarchists (1894), were particularly welcomed by a number of positivist scientific disciplines and conceived as truth-bearers in matters of hereditary individual traits and social behaviours (see Wyndham, 2003: 286). However, the acceptability of Lombroso's assumptions and theories depended principally on their being a sort of scientifically-based systematisation of a variety of previous and contemporary research based on race, class and gender prejudices (see Frigessi, 2003: 102-106; Palano, 2002: 72-73; Giacanelli, 1975: 21). These prejudices were elaborated by Lombroso within a particular and positivistic understanding of the national society, as a "productive and self-sustaining" apparatus disciplined by private and public devices which were functional to the expansion of the "civil society" and capitalistic economy. In his opinion, in fact, – and in those of a large number of European hygienists, sociologists, physical anthropologists, neurologists, and phrenologists – the standard of economical and social productivity, articulated along colour and gender lines, were the most crucial attributes defining the proper citizen. Productivity, decor, social respectability – in sum, the 19th century bourgeois private and public values at the foundation of Victorian society (Strawbridge, 1988: 107-114; Burrow, 1970: 128-136; Young, 1995: 93-98; Shearer,

1996) – together with a racialised and extremely sexist description of individuals were the parameters to discriminate between who could, and who could not, be considered a proper subject and citizen of a new Nation (see Frigessi, 2003: 36-382).

Moreover, intervening in the European debate on races at the time of the first stage of Italian colonial expansion in Eritrea and Ethiopia through a series of statements related to the supposed hereditary inferiority of blacks – and especially Aborigines, located at the bottom stage of human evolution (Lombroso, 1871: 170; Wyndham, 2003: 286) – Lombroso's Darwinist (and monogenist) research turned out to be particularly meaningful in the Western conceptualisation of whiteness versus blackness. In fact, Lombroso's connection between Southerners and black-people, and his deployment of Western categories of race placed his theories on criminal attitudes and racial differences at the centre of the international debate. This allowed them to contribute significantly to the colonial "auto-referential" racialisation – that is, to the process where, in Etienne Balibar's words, "the bearers of the prejudice, exercising physical or symbolic violence, designate themselves as representatives of a superior race" (1988: 39). This auto-referential racialisation was functional, in the broad settler colonialism and in the particular case of Australia at the time of 20th century evolutionist monogenism and polygenism, to the theoretical and practical definition of "what's human" and, subsequently, to the *dehumanisation* of colonized people (Anderson, 2007: 126-145). In fact, even though hetero-referential racialisation in Guillaumin's theory is usually applied to colonial contexts, in the particular case of settler colonialism, where it is not the logic of exploitation but that of the elimination of the "natives" that is the hegemonic principle, the main identity-politics is not hetero-referential but auto-referential (see Veracini, 2006: § 23; Wolfe, 1999: 30). The latter, indeed, assumes that the dominant group is the only human group existing in the colony – having achieved the highest stage in evolution – meanwhile the other "sub-humans" are doomed (or need to be led) to extinction. Very similarly to the Italian "internal colonization" of the Mezzogiorno that, in Lombroso's view, needed to exclude Southerners and reduce the degenerating effects of its brown "natives'" psycho-pathology on Northern civilization, Australian white colonizers – the fittest, and the proper, man – needed to eliminate/exclude/segregate its "black natives". The rhetoric of "degeneration", infertile "miscegenation", and racial decay were deployed in both cases, even if Lombroso referred to them only occasionally – privileging a narrative of "fruitful interbreeding" between Italian "good" Southerners and Northerners and the detention/segregation of "bad" Southerners.

Lombroso's Negrophobia Meets Australian Fantasies of Whiteness

Although formally dismissed after 1936-1937 by the official Fascist doctrine of the unity of the Italian kinship· Lombroso's theories on the inferiority of not-white people and on the criminal attitudes of Southerners, brown people and women continued to hold a great sway on wartime and even post-bellum Italian social and political culture and on foreign scholars.

One of the reasons of the foreign curiosity towards Lombroso was related to his insider viewpoint in discerning and discriminating his own "stock". In those democratic nations like the United States and Australia facing not-so white mass migrations (global "ethnoscapes"), a rigorous and reliable source (transnational "info-scapes", but also "ideoscapes") that would enlighten the dark labyrinth of European racial mixture was needed indeed. A re-articulation of transnational "soma-technoscapes" would help scientists, doctors, politicians and intellectuals in selecting incoming migrants and preserving the white population's symbolical and numerical supremacy. It was in the process of fixing the boundaries of the "purest" and the "fittest" that the supposedly most white turned its gaze to the *other* whites, establishing for them a different status. The mobile line that distinguished those whites was drawn by virtue of a relational definition: the less white exists because the white do, and *vice versa*. And the identity of the former is always externally defined by the supremacy of the latter. In that sense, following Spivak on subalterns (2003: 170) and Fanon on the interdependence of black and white categories (1952: Introduction), one could say that Lombroso, in Italy and abroad, contributed to define, through a clear description of the less/un-fit, not only the "fittest" but the whole symbolic structure of Northern (that is white) privilege.

The first concern, in those countries experiencing mass immigrations from Europe and the Mediterranean, was to preserve an ideal-type white citizen, as testified by the restrictive migration policies approved in U.S. and federal Australia and the plea they made to the Fascist regime – very similarly to what, in late 19th century Australian *gold-rush*, China had been asked for – for a restriction on Italian emigration. In fact, the only way to keep the citizen's whiteness pure was, other than to segregate what was not "commensurable", to reject or reduce who was "potentially the same". To that aim, the work of Lombroso on Italian Southerners (Calabrians, Sicilians and Sardinians) – which locates the white respectively in the North and the less- or not-white in the South of Italy and attributes to the latter an innate and hereditary criminal psychopathology – resulted as very useful. In particular, as argued by Vito Teti, Lombroso's observations on Calabria:

> constituted one of the first psycho-anthropological understanding, based on organicist assumptions and formulated within a positivistic system, applied to the behaviours of the population of a Southern region... In the second half of 19th century, Calabria became a sort of 'laboratory' for the construction of the Southerner's psychology. (Teti, 1993: 152; see Lombroso, 1898: 11, 40. My translation)

Such an account gave rise to a sort of "internal" orientalist imagining (Schneider, 1998: 8) leading to an understanding of Southerners as *alien*. Alienation – as both madness and extraneousness – was conceived of as the main peculiarity of "atavistic", brown (dolichocephalic) Southerners, whose race was not purely white and whose psychology was congenitally criminal. The solution to their criminality was not punishment or detention, but understanding, care, cure and medical segregation (Teti, 1993: 156; Frigessi, 2003: 99-100. Frigessi quotes Lombroso 1868: 91).

In those countries experiencing mass migration, accounts of "degenerating inferior masses able to inoculate the germ of barbarism into a fit and civilized society" (see i.e. Lombroso, 1898: 152-153), fuelled by the vast European literature on domestic subalterns irrupting violently into the public sphere (see also Palano, 2002: 30-44), spawned the horrific, highly racialised image of hordes of barbarians penetrating the virginal white body of the (new) Nation. In these representations, the European trash or the "atavistic populace" – very much like a mob led by primitive instincts and similar in skin-colour and attitudes to the colonial apelike Negro – threatened the symbolic order of the bourgeois society. In the last decades of 19th century, this image of destruction and degeneration (well depicted by Bénédict-Augustin Morel) was explicitly articulated – within the evolutionist reappraisal of traditional anti-black prejudgements – and popularised – through officials, scholars and doctors' massive engagement in social Darwinism – throughout the colonial and post-colonial world.

Anti-black prejudices had a long course before their conceptualisation in terms of biological inferiority, as Windrop Jordan argued in his monumental work on the English, British and American attitudes towards the Black (1977). They referred to an alleged incommensurability between white and black skin-colour; the attribution of ape-like features to Black phenotype and behaviour; and a three centuries-old idea of blacks as the most impure, savage and godless human stock. Since the beginning, these prejudices, within a monogenetic understanding of mankind, in both their broad and situated formulations, crossed British State's boundaries and overseas imperial spaces.

In Australia, anti-black stereotypes, combined with a racial classification that placed Aborigines at the lowest stage of human evolution, resulted in a twofold discourse. On the one hand, it was used to legitimise colonization and colonial genocide towards prior inhabitants. Aboriginal Australians were conceived as not properly *human*: a sort of pre-historical residual of the first stage in human evolution, even lower than African blacks, or as an element of the Australian *flora et fauna* (Giuliani, 2010). On the other hand, it also legitimised white fantasies of a white Australia kept "pure" against not only incoming black and brown, but also *not-so-white* migrants. This was a race-based political anthropology that stripped subjected-to-the-crown Aborigines of the "right to have rights" (Arendt, 1951: 177) and demonised "Asians, Africans, Mediterraneans" as dangerous factors of contagion and decay according to a previous *raciology*. This *raciology*, fuelled by the same race thinking imbibing Lombroso's categorization, was accurately selected and adapted to Australia's race relations in a way consistent with the local needs regarding the Nation-building process.

This twofold discourse was mirrored in the classification distinguishing browns and yellows (not-absorbable migrants but *humans*) from "blackest" blacks (unmasterable *not/sub-human* Aborigines). The *colour-line* between *yellow* and *brown* peoples and *black* peoples marks the passage from a monogenetic to a polygenetic conception of humanity that relinquishes the idea of humans' cultural, social and political "improvability" (typical of the Enlightenment), and embraces instead the idea of improvement as a progressive (biological) whitening. Those who are absorbable in the category of a well-developed humanity (whiteness) are those considered as *whiteable*.

Those who were considered whiteable, in the Australian case, were those people who witnessed the fulfilment of the settlers' first aim: to possess Australia. Indeed, the white Australian "auto-referential" racialisation and its consistent logic of "elimination" distinguished itself from metropolitan societies depicting itself as contaminated from within, as in the case of Nazi Germany. What starkly contrasted with the latter was the Australian answer to the problems of expansion of the settler race and of the resulting "miscegenation": the progressive absorption of half-breeds into the settler category (Wolfe, 1999: 30). This reappraisal of the Australian institutional conception of half-castes in the period between the *Immigration Restriction Act* (1901) and the *Racial Discrimination Act* (1975) helps us to better frame policies like child removal as strategies aiming not to eliminate Aborigines, but primarily to reinforce the idea of a white Australia.

Although that could appear at first sight as the mere substitution of the "white-centred" binary scheme with a "black-centred" paradigm, the articulation and complexity of the theoretical nuances and consequent policies implemented in early federal Australia show the peculiarity of a *colour-line* that is very much fragmented. The two paradigms, the white-centred and the black-centred, work together to build up a taxonomy of mid-colours that compels a continuous re-definition. Both paradigms are engaged in the struggle to define half-castes, where half-bloods are alternately defined as "not-black" or possibly white (Wolfe, 1994: 106-107) or as still-black (Reynolds, 2005: 29-50). Nonetheless, the distinction between black and brown/yellow/less-white draws a geometry that establishes once again the priority of the white-centred paradigm: this geometry finds its own centre in the category of white as self-reflexive, pure and self-consistent nucleus from which all the rest is excluded.

In this sense, one could view the individuation of all those different not-whites in the social space of the nation as prescriptively describing what Australia should be – a white spot. But to admit the possibility of a transformation of the less-white into a (proper, even pure) white "only by not being black" (Mohanram, 2007: xvi; Wolfe, 1999: 30) is to re-admit the black-centred paradigm as a binding one. Nevertheless, in spite of the centrality of this paradigm, the white-centred paradigm continues to fragment the racial spectrum in order to distinguish the "proper" and the "will be" white from the rest.

The distinction between black and brown/yellow/less-white is inherent to a political anthropology that elsewhere I have defined as "of incommensurability", but also "of extraction" of the purest element (or the will-be pure, as in the case of half-castes) from an increasingly "dirty" mixture between inferior and superior human stocks (Giuliani, 2010). Precisely in this sense, in Australia, both exclusion (or even extermination) and inclusion appear as parts of the double process of auto/hetero-racialisation. Inclusion was central, as in the case of half-castes (and absorbable migrants), in order to expand the would-be elected stock (the settlers' and then white Australians' stock). Exclusion/extermination was essential to keep land and home under the white norm. "Inclusion" of half-castes and absorbable migrants, clearly, was crucial to two absolutely different mid-processes: absorption of half-castes so as to neutralize dangerous witnesses of Aboriginal prior occupancy, and absorption of selected immigrants in order to fortify the white numerical supremacy. These processes were leading to the same direction: the creation of a White Australia.

Italiani No Grazie! Lombroso, Cilento and the Racial Architecture of Australia

The early 20th century migration flows were towards the main cities, Melbourne in particular, and the tropical North (Queensland). In the urban context, doctors and hygienists were mostly concerned about a growing and "growingly degenerated" British "white trash" corrupted by the urban way of living.

In the North, public servants and doctors, focused as they were on the white man's struggle for survival, became more and more concerned about his "purity" to transform the prior focus on environmental diseases into a worry about inter-racial contamination and contagion (Bashford, 2004: 110). Interracial contacts became increasingly a matter of great concern, and not only when these regarded white and Aboriginal relations, but also relations between whites and not-so-white migrants. If tropical environmental diseases could be defeated, said the optimist, the purity of the white race had to be defended against external illnesses imported by immigrants and racial degeneration (due to sexual intercourse with the latter).

Amongst scientists and doctors of tropical hygiene, the most strenuous supporters of direct State's intervention on the private and public habits of the northern white population's – with a view to "filling with its healthy offspring and selected migrants the northern empty space" – were Drs. J.W. Gregory, G.L. Wilkinson, C. Cook, J.H.L. Cumpston, H. Sutton, A.G. Price and especially Dr. Raphael Cilento. Cilento had succeeded Dr. Jack Elkington as Director of the Australian Institute of Tropical Medicine in Townsville, and was subsequently nominated Director of the Division of Tropical Hygiene in the Commonwealth Department of Health of Queensland.

Cilento, together with his wife Phyllis, engaged in field research in the Tropics aimed at establishing principles regarding clothing, personal and family hygiene and interracial sexual contact. The explicit aim was to reproduce the "white body", cleansed of the "germs of tropical laziness" (hookworm, but also leprosy), and preserve its purity from diseases and disease-carriers. Hookworm and leprosy were, in their opinion, infecting predominantly Aborigines (traditionally conceived as "reservoir of disease"): the only way to preserve the Nation from degeneration was through Aborigines' coercion and forced containment in lazarettos, measures that were an extension of the Aboriginal camp-system. But the devil was coming also from outside: the white body needed to be protected from incoming diseases and disease-carriers, the so-called foreign earth-eaters, i.e. Pacific "Arabians", Chinese, and Southern Italians (Cilento considered himself as Northerner). In this case, the solution was rejection at the frontier, segregation and quarantine for immigrants, and white peopling of Tropics (Cilento, 1925: 57; see Bashford, 2004: 105-106).

According to a popular *raciology*, Cilento and Cecil Cook described Southern Italians (Sicilians in particular, but also Calabrians and Sardinians) and the other Mediterranean people (Greek and Maltese) as the result of a deadly miscegenation between European, African (*sic*) and Asian (*sic*) races (Anderson, 2002: 159-160; Anderson quotes Cilento, 1925: 241). As such, like other not-white and savage people, they were regarded as generally insubordinate, unstable, undisciplined, too passionate, excitable, loving and hating intensely, inclined to criminality, alcoholism, filth, and sexual diseases.

This *raciology*, focused on an account of Southerners' psychology as "atavistic" and "degenerated", and fully articulated by Lombroso in his research on criminal anthropology· had been very influential in Australia, as witnessed by the great number of policies approved between the 1920s and the 1930s. These ideas found fertile ground in Australian racial thinking of the time – a combination of evolutionism with a positivist idea of race which, like Lombroso's, was untouched by eugenics, but very compatible with the eugenicist idea of *humanness* (Frigessi, 2002: 386, 387; Gibson, 2002: 165).

The success of Cilento's warnings regarding racial purity and of the policies he proposed were strictly dependent on a precise and "active" political anthropology that made the realization of a precise idea of "racial citizenship" its own priority. In this sense, one could say that Cilento's theoretical construction of Australian whiteness can be considered as very similar to the theory of "razza italica" (Italian race) of the Italian doctor and theorist Nicola Pende – a theory that Cilento should have been very familiar to, given his ideological and political proximity to Fascism (Anderson, 2002: 150; Anderson refers to Fisher, 1994). In both theories, the controversy about racial crossing is solved through an idea of nation as crucible of the best stocks – European and Mediterranean in Pende's case, and North-European or Caucasian in Cilento's case. In both cases, this theoretical move consisted in a sort of auto-racialisation (or auto-referred racial identity politics) aiming to define Italy and Australia as properly white nations at a time when the prestige of a (Western) nation depended primarily on its supposed racial unity and undoubted immaculate whiteness. Thus, for theorists like Lombroso, Cilento and Pende, to be proud nationalists meant to be proudly racist and engage in racial hygiene and engineering.

In Italy, Pende's hypostatisation of such a *razza mediterranea* (Mediterranean race) or *italica*, as the result of the fruitful encroachment of different *stirpi italiche* (Italic kinships) (1933: 234) "embodying the ancient qualities of the Roman Republic and Empire" (Pende, 1933: 213; and broadly 212-221), was meant to satisfy a double need. On one hand, it came to terms with the traditional localisms and the division between North and South. On the other, it built a symbolic construction of unity essential to the totalitarian idea of an organic nation. The "razza italica", wrote Pende quoting Mussolini,

> is and needs to remain Mediterranean; in thousands of years, indeed, regardless of its vast crossbreeding with a number of other races' kinships (*stirpi*), it maintains prevalently the ancient Italics' blood, which constituted, since prehistoric times, the anthropological ground of the Peninsula. (1933: 227; see also 212)

In this sense, the traditional anti-Southernism scientifically expressed by Lombroso had to be set apart so as to privilege a narrative of unity. Thus, wrote Pende quoting Mussolini, "*da felici mescolanze sono spesso derivate la forza e la bellezza di una Nazione*" (1933: 227). This rhetoric of the "fruitful interbreed" was a narrative ingrained in a Catholic background and a broad refusal of eugenics. Despite Lombroso himself hoping for an increased intermixing amongst the various Italian stocks – Nordic or Arian (North), Greek-Roman, Semitic and Hamitic (South), or in other words, *dolichocephalic* and *brachycephalic* (Lombroso, 1898: 53-57, 81) – in order to homogenize Northern and Southern kinships, and thus emancipate the latter from its genetic inferiority (Lombroso, 1876: Vol. III; Pende, 1933: 240), his striking condemnation of the Southerners was not endorsable by the regime and its racial identity-politics.

Such a symbolic construction of unity, in Cilento's opinion, was not the core of Australian concerns, which should rather be focused on racially engineering a unique national race synthesising the British stock and other higher stocks, and excluding or even neutralizing (through sterilization or permanent clinical detention) unproductive stocks. Accordingly, the Australian racial building-process could be viewed as an unaccomplished exercise in engineering a topical idea of racial unity and homogeneity. Indeed, the main concern was not to include something already existing in a unique strand, but to exclude few still existing racial components and discourage not-so-white migrants from filling the national homeland (as confirmed by the 1922 *Australian Settlement Act*). The result of this racial engineering was to be a white Australia consisting of an optimum racial mixture of British and Nordic (or Nordic and Alpinian or Caucasian stocks, depending on the view-point) that included increasingly whitened half-castes. In order to obtain the greatest result what was needed was the (racial) exclusion of Mediterranean stocks, the containment of the "yellow peril", and the confinement of Aborigines, as well as good personal hygiene, medical care, quarantines and disinfestation. In Cilento's words:

> Queensland was the first (and for long time the only) instance of successful colonization of a tropical and subtropical land by a population almost entirely white... beginning more or less as an accident of location, the process has been continued by design with the aid of three significant factors: (a) the absence of any teeming native coloured population riddled by endemic diseases; (b) the increasing capacity to control introduced diseases and to develop a high standard of living; and (c) the constantly increasing proportion of the population that is Queensland born. (Cilento, 1959: 421)

Walking with the Living Ancestor: towards a new Understanding of Multiculturalism

As we have seen, Cilento shared with many doctors the idea that Aborigines were to be "segregated" and "distanced" from white citizens to prevent them from spreading diseases and threatening the Australian way of life. Aborigines were damaging to whites as well as to themselves, for their presumed lack of self-control, discipline and respect for basic hygienic measures. Nevertheless, they had to be "preserved" from extinction: in a growing opinion, they were, in fact, increasingly regarded as the last sample of "archaic" or "Stone Age" Caucasians (Cilento, 1959: 177-191) and as such, both the "object" of public hygiene and matter of archaeological and scientific interest.

In Cilento's view, Aborigines, just as "atavistic" Southern natural-born criminals or Negroes in Lombroso's (1871: 171), needed to be kept apart from the society in order to reduce chances of microbial contagion and genetic exchange and to let them live as examples of the evolutional stages of mankind. Yet, if Southern natural-born criminals were conceived of naturally un-eliminable – as a self-reproducing congenitally un-absorbable social group (Lombroso, 1868: 6-7, 89-91; 1878) – Australian Aborigines needed to be sanitized so as to be preserved from the devastating encounter with civilization.

Hence, this particular evolutionist understanding of Aborigines – just as positivism in Italy and Europe – re-admitted who had been outcast "natives" into both mankind and the *body politic* but in a renewed regime of "protection", "segregation" and "dependence". Following Cilento and other commentators' segregated Aborigines, as "archaic" (pure) or "dark" Caucasians, would be reduced to an unanimated object of national welfare and scientific research. Half-castes, as the result of the encounter between "distant relatives and their contemporary offspring", would accordingly be forcibly included in the nation as "becoming-white" citizens (Anderson, 2002: 193-194).

This new conception of Aborigines, formalized in Australia after the Second World War and in particular after the 1970s, led to the reduction of Indigeneity to a mere "social component" of the new nation. This is a conception that takes uncritically the history of Australian white colonization and its violence, and that equates the problem of dealing with Aboriginal presence to the problem of containing and separating the new immigrants. Aboriginality, thus, was finally included in a all-encompassing racial geometry formalised by Australia's immigration policies, and, as Joseph Pugliese put it, "structured by hierarchies of ethnicity which operate along often unstated but functional lines of racialised/racist ranking – from the 'potentially-the-same' to the outer limits of the 'absolute other'" (Pugliese, 1995: 241).

This is a hierarchy of "races" – or "ethnicities", or "colours" – that finds its political subject and its governmental object hypostatised by that racialised thinking forged in the 19th and 20th centuries, by virtue of the intertwining of evolutionism and eugenics. This hierarchy maintains at its core the assumption that Australia has always been a white nation (by virtue of an auto-referential racialisation) and that its "others" derived their diversity – as the result of both racialised auto- and hetero-referential identity-politics – from their racial/spatial positioning in relation to a supposed originally white body/space.

Alekos Doukas (1900-1962)

A Dis-Located Life in the Shifting Terrain of the Eastern Mediterranean[1]

Petro ALEXIOU

Affiliated Fellow, PhD, at Macquarie University, Sydney

Introduction

This chapter aims to delineate an approach to the Eastern Mediterranean as a location for intellectual and ideological movements, as well as historical events, that impinged on, and to a large extent, shaped the early life and thinking of an historical subject, Alekos Doukas who emigrated to Australia from Greece in 1927.[2] It is an integral part of a broader study of a writer and political activist that attempts to disrupt and complicate the assimilationist construction of non-Anglo migrants in Australian history. It does this by framing the subject in a transnational context and by exploring prior discourses and power regimes in his life. The exploration of discursive history as the embodied experience of an historical actor has the potential to question many common assumptions about the past and probe the silences and ellipses in national and collective narratives.

The biographical approach, while interdisciplinary, is located theoretically within a poststructural discipline in which biography is practiced as an integral part of cultural history – the focus being on the located subject, embodied history and archival specifics, but of equal importance, the individual subject within institutional and discursive history. As David Carter (1989: 35) has succinctly put it:

> Biography as cultural history is the story of the individual, the "biographised" subject, known within and across... institutions and discourses. It

1 For a more detailed and comprehensive account of the topic of this chapter see Alexiou (2008). All quotations from Greek texts are my own translations.

2 He is the youngest brother of the well-known Greek writer and critic Stratis Doukas (1895-1983) with whom he corresponded intimately for most of his life.

thus might also be the story of social institutions, and of ethical, political, and aesthetic discourses, known within and across the biographised subject.

The title of the chapter suggests a number of important concepts: location and dislocation, the chronotopic coordinates of specific time and place, the presence of disturbance and change, as well as the individual life within broader historical events and discourses. In this latter interrelationship, the focus is on the "micro-political" level in which the subject exercises agency in the cultural production of everyday life (Laliotou, 2004: 13). It is here that the study concentrates much of its attention in order to observe how broader discourses manifest in the personal sphere and how at certain times individual thinking undergoes important shifts and transformations. It takes up Michel de Certeau's (1984) idea of the "singularity" of the ordinary historical subject and the polemical interplay between discursive power and the "trajectories" of everyday practice.

In the relationship of the micro and macro in history, the Eastern Mediterranean of the early 20th century figures as a space of competing forces and discourses that at critical times intersect or converge. These points are important in the tracking of a subject through whom historical movements often manifest in singular and contradictory ways. My aim is to sketch some of the major historical differentiations that comprise these larger movements and which in turn illuminate shades of meaning and complexities in a narrative of the life of a refugee intellectual who migrates to Australia. After a short biographical portrait that includes a 'fast forward' to Doukas as he has appeared in Greek-Australian studies in recent decades, the chapter sketches key periods in his life up to the moment he enters the port of Fremantle in November 1927. This exercise in discursive history and biographical narrative may help us grasp the apparent contradictions and paradoxes that mark the entry point of such a subject into White Australia in the late 1920s, another important intersection of tangential but related discourses of a dominant white colonial society and the Greek diaspora in the antipodes.[3]

A Brief Biographical Portrait

Doukas grew up on the island of Moschonisia (Turkish *Cunda*) and the mainland town of Aivali (Turkish *Ayvalık*) on the Asia Minor coast in the Ottoman Empire. He left for Greece as a fourteen-year-old in a period of exile during World War One, returning in 1919 after which he fought and was wounded in the Greek-Turkish War (1919-1922). He

[3] For an analysis and account of Doukas' first ten months in Australia see Alexiou (2005).

was evacuated to Greece where he spent five years as a "refugee" primarily in the "New Lands" of Northern Greece.[4] In Australia, between the years 1927-1937, apart from a return to Greece 1932-1935, Doukas worked as a seasonal fruit picker and in Greek cafés and fish shops which by this time constituted a network and enclave of an ethnic catering industry developed along traditional diasporic patterns from the 1890s. Although Doukas did pursue his writing in the pre-World War Two period (an unpublished romantic novella, serialised articles on Australian history, prose poems), he only became well-known in the 1950s for his novel *To Struggle, To Youth* (1953) and the posthumously published *Under Foreign Skies* (1963). In the 1940s and 1950s, he translated literary works into Greek from English and co-founded an important leftwing Greek-language magazine and a newspaper in 1951 and 1957 respectively.[5] He became politically active during World War Two as a communist, his activism primarily channeled into writing and translation. In the late 1940s and the 1950s, he was active in the international organisation League for Democracy in Greece, and advocated for the rights of Greek political prisoners.

In brief, Doukas' thinking underwent a process of radicalisation during the 1930s, due primarily to the political crisis in Greece and his experience of the Depression in Australia. From an introverted young Anatolian refugee recuperating from wartime traumas and searching for spiritual and metaphysical solace, he became the intellectual representative of the growing Greek immigrant left in Australia during World War Two and after. Following his sudden death in October 1962 from fatal injuries received after being struck by a car, Doukas' life and work took on certain legendary qualities amongst Greek migrants, in particular on the left side of politics (Alexiou, 2009). Through a process of oral and written mediation, he has been invariably represented as a unitary socialist subject who encapsulates the radical consciousness of a migrant collectivity during its early years of settlement (Tzoumacas, 2003: i). While this interpretation of the past may satisfy the need for a culturally progressive narrative of a migrant group's identity, it fails to grapple with the complexities of historical evidence. Since the 1980s, in a period in which multiculturalism has been influential as a literary and cultural discourse, migrant literature has often been read as a testimonial and biographical genre rather than constructed and imaginative literature (Gunew 1994). Doukas' two postwar 'historical' novels, examples of an

4 For accounts of the role of refugees in the Hellenisation of Northern Greece see Kontogiorgi (1992); Yerolympos (2003); Mackridge (1997).

5 The magazine *Ellino-Afstraliani Epitheorisi* (Greek-Australian Review) 1951-1953, which was systematically denied a publishing permit, and the newspaper *Neos Kosmos*, the largest circulating Greek newspaper in Australia today.

immigrant Greek literary engagement with Australian postwar socialist realism, have been widely read as unproblematic testimony rather than fiction, which has led to a view of him as a "pre-literary" witness of inter-war migrant history.[6]

In the broader field of Australian cultural history, Doukas is virtually unknown, not so much as an empirical deficit but more importantly as the absence of a discursive space for the theorisation of such a figure. To dismiss the absence as simply a question of language (in)accessibility is to undoubtedly diminish the significance of the historically powerful disciplinary regime of assimilationism that has rendered non-Anglo-Celtic and Indigenous cultural life illegitimate foreign currency in the dominant cultural economy. In an Anglo nationalist and assimilationist framework, the migrant subject is constructed in terms of his (rarely her) integration into the hegemonic cultural norms of the country.[17] Other aspects of the migrant subject have been considered extraneous to the process. In the dominant racialised discourses of society in the early 20th century, the Southern European is a subjugated figure. In the discourse of the Greek communities, a split subjectivity can be discerned in which diaspora national loyalty and identity were retained but subordinated within a broader submission to White Australia and its colonialist foundational myths (Alexiou 2005). This historical process, full of contradictions and conflicts, is usually treated to a smooth assimilationist reading. My study of a located historicised subject aims to explore these areas of contradiction and move beyond a homogenised view of migrants as objects of knowledge. In this approach it is crucial to study the processes of subject formation prior to the migrant's arrival. A part of the research equation involves the unpacking of the subject's cultural baggage, the discursive systems bearing on the individual and the collective *before* as well as *after* migration. In this chapter my focus is on the *before*, as a beginning point for an understanding of the *after*.

In Alekos Doukas' life we have a biographical narrative that crosses national, imperial and numerous community histories. In Australian history such a life, if acknowledged at all, has been largely flattened to fit a monocultural account, whereas in Greek-Australian studies it has been somewhat mythologised within a working-class migrant narrative. My exploration of transnational Mediterranean relations is an attempt to question such constructions and suggest more complex and messy terrains.

---

6 For a broader discussion see Carter (1985, 2002).

7 This premise concurs with Laliotou's (2004) broad theoretical conceptualisations in her study of Greeks in the U.S.

Where to Begin?

A study that attempts to write biography as cultural history is confronted with the task of articulating "beginnings", not as pure origins but as an analytical and historical delimitation of an inherently large archival body (Said 1985). In both a biographical and biological sense, a life may be said to have no precise beginning, at least as a non-arbitrary point in time. Similarly, broad historical movements or discourses may also be said to begin only in the arbitrary act of analysis and writing. The narration, however, of a life story invariably assumes a beginning, middle and end. This is of course a generic imposition, and as Hayden White (1987: 23-25) has noted, it represents the writer's own moralising stance towards the past and present.

With this in mind, I attempt to construct a life story out of discrete texts and heterogeneous genres: letters, fragmented notes, published and unpublished literary works, 'historical' or semi-autobiographical novels, reminiscences and oral testimonies.[8] These texts, as well as some artefacts,[9] are used as ways to "figure forth the thought-world" of the biographical subject and "the world of emotional investment and praxis of [their] time and place of production" (White, 1987: 187). Even so, without a more systematic representation of wider societal discourses and foundational myths, such texts are incomplete, fragments without context. The study therefore aims to provide the wider context of societal discourses, foundational myths and broad social formations that operated in the Greek world of the Eastern Mediterranean at the turn of the nineteenth and twentieth centuries. At the risk of oversimplifying, I will sketch their broad contours, focusing on the complexities and differentiations that existed within and between Greek populations. The historical context is that of growing and powerful national/nationalistic discourses that narrate the past as a self-evident course of events. The interpretation of the past through the "logic of the nation-state" is, without doubt, a pervasive conceptual obstacle in such an undertaking (Kitromilides, 2003: 28-29).

To begin with, we are dealing with economically and socially semi-autonomous Greek communities in the Ottoman Empire, itself a multi-ethnic and religiously-defined regime seriously weakened by the penetration of Western economic and strategic interests in the Near East. Within this empire, a Greek national irredentist movement simmered,

8 Most of Alekos Doukas' letters and unpublished manuscripts are located in the Stratis Doukas Archive, Department of Medieval and Modern Greek Studies, Aristotle University, Thessaloniki.

9 I refer to the extant bullet that wounded him on 13 August 1922 and his war medal, the Silver Cross of Valour, bestowed on him by the King at the Front on 8 July 1921.

stoked by the policies and politics of the neighbouring Greek state. The Greek populations in the empire were by no means homogenous in language, history or social organisation, but were nevertheless increasingly drawn towards a Greek national identity through the activities of a network of cultural and educational associations (Exertzoglou, 2000). The nationalist discourse of Hellenism was displacing an older practice and adherence to Ottomanism (the ideology of empire power) although the tension between them was often expressed in ambivalent ways.

Here we need to differentiate the Ottoman Greeks from the Greek diaspora trading communities of the Eastern Mediterranean which, although linked to the former through family, business and inter-migration, were located outside the empire and had different relationships with local regimes.[10] The Greek diaspora communities in the modern period stretched back at least to the 18th century and played a decisive role in the ideological and material preparation for the founding of the Greek nation-state in 1832 (Kitroeff, 1991). We need to keep in mind however that the traditional or "mobile" (the so-called "archetypal"[11]) diasporas such as the Greeks had historically based themselves in multi-ethnic empires as a class who provided communication networks and specialised roles as merchants and diplomats in a reciprocal exchange relationship with the ruling elites (Armstrong, 1976). Despite their ideological investment in the Greek state, the Greek diaspora still maintained a certain supra-national or supra-territorial identity sustained through ethnic myths that were not necessarily tied to a nation-state.

In both the Ottoman Empire and the diaspora proper there were impulses and movements resistant to nationalism. In the former, irredentist ideology, expressed historically as the "Great Idea", was not always enthusiastically embraced for reasons of self-interest and long-term survival. While the growing number of Greek educational texts, histories and geographies in particular, promoted Greek national consciousness through a "unity of time" and "space" (Anestidis, 1989: 18) it is important to note the existence of a simultaneous and overlapping Greek Ottoman intellectual outlook that differentiated the Greek minority interests from those of the Greek state (Kechriotis, 2005). Similarly in the diaspora centres, political encroachment by the Greek state in community affairs was not always welcomed by the Church or unanimously supported by the local community elites. Even so, by the early

[10] For an account of the Greek diaspora communities of Egypt see Kitroeff (1989).

[11] The concept of an "archetypal" diaspora, used loosely by Armstrong (1976) and postulated by Safran (1991), has been critiqued as an "exclusivist paradigm" that as a prescriptive definition is unworkable (Clifford 1997: 247). The comparative approach is nevertheless valuable as a way "to specify a complex discursive and historical field" (Clifford 1997: 249).

20th century the nationalist ideology of the "national centre" (the Greek state) had begun to impose its dominance in both spheres.

The key to an understanding of the modern Greek diaspora experience hinges on locating the historical period of transition that occurred between the traditional diaspora settlements and the modern labour migration or "proletarian" diaspora which roughly corresponds to the large wave of immigrants that migrated across the Atlantic to the U.S. in the period 1890-1924 (Laliotou 2004: 53; Kitroeff, 2002: 362-363). In the interwar years, these changes led to considerable indeterminacy and overlapping in the discourses related to new migration and older diaspora communities. In addition, the Greek defeat in the Asia Minor Campaign of 1919-22 not only resulted in the collapse of the Great Idea and irredentist nationalism, it also changed the dynamics of Greek homeland-diaspora relations (Kitroeff, 1991). This landmark event would have long-term repercussions for Greek communities in Australia.

Drawing back to a wider framework, we note that Greek nationalism, or Modern Hellenism in the Near East, is profoundly influenced by Western notions of progress and expansion. These notions are implicated in Orientalist discourses such that Westernisation is internalised and redirected by Greeks towards the Oriental *Other*. It is *in* this mix of intersecting and converging discourses that the subject of this study is located at a time of competing nationalisms and expanding Western presence that led to genocidal conflicts, war, uprooting and the redrawing of borders and identities.

Constructing a Childhood

In narrating Doukas' childhood I situate the available biographical information and texts within the discursive space of the educated Ottoman Greeks' self-definition as a community though time and place. By studying the historiographic tradition of local history writing, we begin to grasp the foundational myths of the neighbouring Greek towns of Aivali and Moschonisia. These narratives can be read at various levels, their contradictions or "hidden meanings"[12] pointing to important shifts through time, for example, the subtle shifts in meaning in the local histories between "pre-nationalist" and nationalist consciousness, and the narrow and broader meaning of "*patrida*" (country). In short, they reflect the very mercurial nature of an empire minority caught between displays of Ottoman loyalty and nationalist or separatist consciousness.

---

[12] Such contradictory aspects arise through the "'dilemmatic' nature of language" which requires that we pay attention to "hidden meanings", in discourse (Parker, 1992: 15, 23).

The biographical details of Alekos Doukas' childhood years can be summarised easily enough. The youngest child of a couple whose families belonged to the island's notables, he followed in the footsteps of his three older brothers through the local primary school to the highly-reputed High School of Kydonies in Aivali on the mainland where he enrolled in 1913. Although the school was forced to close in 1915, he managed to absorb something of its unique liberal education which was historically linked to the famous Academy of Kydonies, a renowned centre of learning during the Modern Greek Enlightenment (Clogg, 1972). Doukas' family were landowning merchants with extensive olive groves, olive-pit factories and oil presses, but who experienced bankruptcy in 1912. The security of young Alekos' world had begun to falter even earlier, when, following an initial optimistic response by Greeks to the promises of the 1908 Constitution, the political tide turned against them as a minority. In June 1909, martial law was declared, and Aivali and Moschonisia were blockaded by land and sea (Sakkaris, 1922: 202-206; Karamblias, 1949: Vol. II, 212). In the next six years the clouds on the horizon grew darker as the events of the Balkan Wars, conscription and World War One impacted on the towns. In 1915, Doukas' childhood suffered a rupture in two senses: his father died and he was sent to the nearby island of Lesvos, only recently incorporated into Greece, to continue his education. A few months later, fleeing a forced evacuation of the island by Turkish troops, his family followed. Thus began the Doukas family's first exile.

The biographical details above are the bare bones of a childhood narrative; missing is the wider web of societal discourses that imbued the life of Alekos Doukas and his contemporaries. The educated discourse of the two towns, manifest in a growing number of local histories, is preoccupied with foundational narratives, communal identification with land/landscape and historical continuity, the latter invariably traced back to ancient times.[13] The theme of the cyclical fortunes of the towns, a pattern of exile and return, is deeply ingrained as an organising principle and relates to deep binary notions (catastrophe-rebirth, destruction-rebuilding, stagnation-prosperity, inertia-progress, fall-revival, barbarity-civilisation, ignorance-learning) that surface at numerous levels. This mode of perceiving historical process has its roots in the European Renaissance and was further revived in the period of the Enlightenment when Greeks were constructing narratives of an emerging nation-state (Liakos, 2002: 28-30). The deeply embedded belief in

[13] The most important sources are Drakos (1895); Apostolakis (1914); Sakkaris (1920); Karamblias (1949).

the cycle of suffering and redemption appears to be inextricably intertwined with the ideology of nationalist irredentism (Patriarca, 2005).

The local Greek histories are, however, not only about connection to country in a local sense, they are inextricably connected to wider Greek historiography which amalgamates classical heritage, Christian Orthodoxy and the modern nation-state into a cohesive national narrative in which the local Greek populations of the Ottoman Empire can locate themselves in historical time and space (Petropoulou, 1982). Educated discourse is also connected to the 19th century national project of Folklore Studies (*Laografia*), underpinned ideologically by an assumption of the nation (the *ethnos*) as an "eternal verity" (Herzfeld, 1986: 13). The discursive connections are at once local, regional, national, but also transnational – Western influence through Orientalist discourses in which the Greek presence in Asia Minor is associated with entrepreneurial activities and civic progress. Greek education of the time is the *sine qua non* of progress and a particularly strong feature of the twin towns in which Doukas grew up. Reading his fictional reconstruction of his early life, one is struck by the degree to which the contemporary discourses of his childhood constitute the deeper structures of its concepttual framework (Doukas, 1953). Despite his 1950s Marxist critique, Doukas' memory is preoccupied with the themes and concepts of local historiography and folklore. Landscape, ancient history and folklore dominate.

Doukas' fictionalised account of his childhood years is also connected to a contemporary post-World War Two Greek discourse of the "lost homelands" which foregrounds a particular set of memories, events, betrayals and hopes, all issuing from the landmark event known as the "Catastrophe" of 1922 (Liakos, 1998). In this discourse of memory, Aivali commonly occupies a pristine cultural space where Greeks of the Ottoman period enjoyed a high level of prosperity and cultural wholeness that was subsequently lost.[14] The "lost homelands" discourse undoubtedly constitutes a complex set of responses to the past that are not unrelated to broader problems of modernity and displacement (Papailias, 2005: 60-63).

By invoking and exploring these discourses, we begin to understand Doukas' life, not as unmediated or privileged insight, but as the structures, categories and textual modes through which he himself understood his life retrospectively. That he disagrees in the historical intro-

[14] A good illustration of this is the popular Greek song of the 1990s "*Ta Smirneika Tragoudia*" (The Smyrna Songs). See CD *Astranammata* (lyrics Ilias Katsoulis, music Pantelis Thalassinos) 1996, The Music Box International. See also Kostis Moskof (1982: xvi).

duction of his 1953 novel on certain matters of ancient history with older writers such as the Moschonisiot, Efstathios Drakos, is less significant than the fact that, like other historians of the towns, he considers the interpretation of the ancient texts the foundational starting point in the discursive reconstruction of the "lost homeland". In Pierre Bourdieu's (1971: 191) terms, Doukas shares with his contemporaries a certain "*habitus*" where differences can be understood through their deeper commonalities.

Doukas' childhood years end in a period of armed nationalist struggles in the Balkans, harassment of minorities, refugee movements (Muslim Turks fleeing into a shrinking empire, and Greeks and Armenians escaping the other way) and World War One which exacerbates all of these. While these larger events are not the primary focus of this study, they stand as narrative divisions, landmarks in the personal and social landscape, and sometimes, as physical forces impinging on the body of the subject. They may signify collective experiences, but under closer scrutiny individual subjects experience them in singular ways.

Location resonates in particular ways in this study. Doukas' birthplace is only 80 kilometres south-east of the historic site of Troy, and 160 kilometres south of Gallipoli,[15] the site of the Australian foundational national myth of the Anzacs.[16] The Dardanelles is a point of historical crossings between the notional East and West – Xerxes' invasion from the east, Alexander's from the west. These historical connections intimate the contention that surrounds the naming of place, the significance of geographic sites in national myths, the geographic markers in Orientalist narratives, and the contemporary dimensions of such a study. To ignore them would be to reproduce a hermetically-sealed national or 'ethnic' biographical narrative of an Asia Minor Greek subject that ignores the entangled histories of this specific geographic and cultural space that was once the communities of Moschonisia and Aivali.

A transnational perspective offers the further possibility of juxtaposing Australian and Greek histories. The experiences of Alekos Doukas, his family and the island community of Moschonisia, and the Australian troops deployed in the Dardanelles in 1915 occurred in close geographic proximity but as narrated histories are never connected, as though they

[15] The Australian name "Gallipoli" is the phonetic amalgam of the Greek *Kallipoli(s)* and the Turkish *Gelibolu*.

[16] Australian and New Zealand Army Corps. Now celebrated as "Anzac Day", an annual national celebration on 25 April.

belong to different discursive universes.[17] The Australian Anzac story does, however, have connections to the populations on the Asia Minor coast, both in a synchronic and diachronic sense. As an Orientalist construction in which the classical Greek warrior narrative is reconfigured as another Western confirmation of manhood/nationhood forged through heroic struggle against an Oriental horde, it is not unrelated to the intersecting discourses that operated in the Greek context at the end of World War One

The March to War

In the period 1915-1922, the conjunction of discourses and events have critical consequences for both Doukas and the populations in this part of the Mediterranean – world war, blockades, forced movements of civilians, exile, collapse of the Ottoman state, occupation of Turkey by Greek forces, victory of the Turkish republican forces, exodus of Greek civilians and finally the compulsory exchange of Christian and Muslim populations in 1923,[18] an exercise that Lord Curzon cynically dubbed "unmixing peoples" (Hirschon, 2003: 4). How do we understand an individual historical actor in such a complex conjunction? Doukas' thinking and choices can be understood within the discursive spaces of specific Greek sites associated with his life and, by broadening the focus, the powerful political discourses that were clamouring for particular solutions to the "Eastern Question" at the end of World War One.

Lesvos (or Mytilini[19]), where Doukas spent a year at the High School, was a significant cultural and intellectual centre. With a developed local economy and financial and social institutions, it had become by 1912 one of the main trading gateways into the Ottoman Empire (Sifnaiou, 1996). The island was an early centre of Greek socialist ideas and the site of the demotic literary movement "Lesvian Spring" which produced a significant number of Greek writers and artists, including Alekos' brother Stratis (S. Doukas, 1976: 38-52). The island was a

[17] It is only in recent times that Australian writers have begun to contemplate the proximity and interrelationship of the Australian participation in the Dardanelles Campaign in 1915 and the Armenian Genocide. This historical silence also applies to the history of the Greeks on the Asia Minor coast. See Manne (2007).

[18] "Convention Concerning the Exchange of Greek and Turkish Populations, and Protocol, Signed at Lausanne, January 30, 1923". The exchange involved about 1.5 million people which included those who had left Turkey during the war or immediately following the Greek defeat, while the rest, both Christians and Muslims on each side, departed in more regulated circumstances between 1923-1926. See Hirschon (2003).

[19] The capital and port, Mytilini, is also the name often used for the island of Lesvos (also Lesbos).

stronghold of republicanism led by Eleftherios Venizelos, who was regarded, especially by the recent refugees from Turkey, as a charismatic liberator who would continue the irredentist successes of 1912. Although Doukas does not write extensively about this period it is clear he absorbed the demoticist literary and Venizelist outlook.

In August 1916, Doukas joined his family in Thessaloniki, capital of the newly-incorporated northern territories, and by September, seat of the new provisional republican government that was the culmination of a national political schism (Petridis, 1997: 186-188). Venizelos' provisional government aligned itself with the Anglo-French Entente and entered the Balkan theatre of war with the view to furthering Greece's irredentist claims (Clogg, 1992: 83-97). With the family shift to Thessaloniki, Doukas' schooling ended and he began employment as a clerk with an English stevedoring company in the port, while his brother Stratis enlisted in Venizelos' "National Defence" forces. Like Lesvos, Thessaloniki was a mix of old and new, its multi-ethnic Ottoman character undergoing a process of Hellenisation in a period of war, refugee influx and the presence of a large number of British and French troops using the city as a supply base.[20] In Alekos' accounts, there are no specific memories of the city or events of the time, including the Great Fire of August 1917 that essentially destroyed old Thessaloniki and the predominantly Jewish commercial section (Hekimoglou, 1997: 200). That there were dark undercurrents of nationalist chauvinism and anti-Semiticism at the time is undeniable, although this has been silent in Greek historiography until recently (Mackridge, 1997; Avdela, 1998; Liakos, 1988; Mazower, 2004). Doukas' postwar fiction displays a similar blind spot, reflecting this silence and demonstrating how events can be selectively self-censored in the memory of historical actors. The end of World War One brought with it Turkey's surrender to the Entente and signalled a turn in fortune for Asia Minor refugees.

Doukas' return to Moschonisia in the spring of 1919 coincided with the occupation by the Greek army of the Smyrna zone which extended as far north as Aivali. The occupation was welcomed by the Ottoman Greek communities as a liberation although there was an undercurrent of disquiet about the possible turn of events. It signalled a reversal of power in inter-communal relations. The Greeks, formerly a vulnerable minority subjected to harassment and internal deportation during the war, now had the upper hand. Jingoistic nationalism dominated, as

20 By January 1916 almost one quarter of a million French and English soldiers had already disembarked in the city. See Petridis (1997: 186-188) and Mackridge (1997: 10,13).

thousands of young men from Aivali and Moschonisia enlisted with the Greek army.[21]

It is important to widen our frame of reference here and consider the conjunction in this period of Greece's nationalist irredentist movement in relation to the Greeks of Turkey, the interests of the European Powers in the defeated Ottoman Empire and the dominant Orientalist discourses of the time. With the occupation of Asia Minor, Greece was finally to put to the test a messianic dream of enlarging its territories to encompass those of the Byzantine Empire.[22] This involved a serious engagement with the *realpolitik* of the Great Powers who were pursuing their own geo-political interests. "The Eastern Question" – essentially a political discussion about Western interests in the East – was by definition an Orientalist discourse. The influential Greek diaspora of England, for example, called on European support for Greece's ambitions by enumerating the essentialist features of Turks as incorrigible Asiatics, barbarous and despotic, and arguing that Europe owed Greeks a debt for their classical heritage and that a powerful Greece was in the interests of the West (London Committee of Unredeemed Greeks, 1919a & b; Polybus 1919).

The conjunction between Greek and European discourses is clearly illustrated in a talk given in Athens in April 1918 by Gustave Fougéres,[23] the Director of the French Institute and President of the Greek-French Association, pointedly titled "The Mission of Hellenism and the Greek Soldier's Duty". Fougéres' broad conception of history hinges on the idea of "the age-old conflict of Europe and Asia", evoked with the image of historical scenes on a bas-relief, depicting in processional manner, a three thousand year-old continuous struggle (Fougéres 1918: 9). In a statement of deep Orientalist logic, Fougéres informs his audience that between these "two contradictory worlds. No reconciliation is possible" (10).[24] He presents a view of Asia Minor as seen from Greece, a perspective reminiscent of the Greek term *i kath' imas Anatoli* (Our East), which, whilst constituting a moral order, also provides an economic subtext of a vast space that offers "under-exploited and inex-

21 Ralli (1995: 91), without giving a source for her figure, writes that over 4,000 young men joined up to the "call of their sweet motherland".

22 See the popular map of 1920 which shows Venizelos' "Greater Greece" (M. L. Smith, 1998: Front cover). Kechriotis (2005: 27) argues that the Great Idea was a shifting concept associated with "overlapping national discourses" and that its final phase of empire ambition was based as much on Greek-Ottomanism as it was on the older "nation-state irredentism".

23 Fougéres was a classical scholar with archaeological experience in Greece, professorship in France in History and Greek Philology, and a resident of Greece from 1913.

24 See Said (1978: 149-166).

haustible treasures, the gold of its rivers, precious stones, grains, fruit, flocks, horses and the variety of cattle [that] make it the Promised Land" (10).

Fougéres, the prestigious scholar, thus delivers an argument to his Greek audience that is a vigorous endorsement of the "Great Idea". This is the "struggle today", the duty of the country and its soldiers, who having valiantly accounted for themselves as Allies in World War One, must soon proceed to liberate their countrymen (26). For Fougéres, and his audience both near and far, the "educational power of history" is unambiguous (5). It exhorts the taking up of arms against a hereditary enemy of the West and Greece's glorious heritage. The speaker ends by describing how, as he watched the Greek soldiers parade on Independence Day the previous month, he had a premonition of victory. A voice within him spoke: "Oh Greek soldiers, you are truly manly. I don't ask your names, for I know them already. Your names are surely Achilles, Ajax, Diomedes, Leonidas…" (30). Just as the East's corruption and evil is known in advance by the Orientalist so too are the virtues and strengths of the West. Fougéres' speech in April 1918 is a call for war and an argument for enlistment, "the Greek Soldier's Duty" to defend "the eternal Greek ideal with your life even" (30).

Let us zoom in on our subject once again. The fact that Doukas worked at the island's Municipal Offices and tutored school students for ten months before his enlistment in the Greek army in February 1920 suggests that he did not come to his decision quickly. The decision is illuminated by a brief account he later wrote to Stratis in a letter. This micro-political event is worth examining as it reveals how a subject might reproduce or apply "interiorised master patterns" of the educational "*habitus*" (Bourdieu, 1971: 192-193). The letter refers to the period before Alekos' enlistment and his crisis of conscience is significantly located in an educational context:

> I remember that before I enlisted in the army… I was tutoring on the island. Then, while teaching the children, the scene between Socrates and his disciple came up in the reader. The disciple went to Socrates at night to help him escape. The scene of the Laws. At that point I was severely tested. I heard the harsh words hanging over my head. "Why aren't you enlisting?" And I didn't escape from their grip until I enlisted as a volunteer.[25]

The "scene of the Laws" from Plato's *Crito* is an important text in Western political philosophy, dealing it as it does with moral and political issues of duty and obedience to the state. Plato introduces the personified Laws in a dialogue within a dialogue in which they speak to

---

[25] Letter 28.10.1924. *Op. cit.*

Socrates regarding his obligation to respect the state's decision to execute him. The Laws introduce a parallel argument of obligation based on a parent-state analogy, duty to one's parents implying duty to the state. The key passage may well have been the following:

> if you cannot persuade your country you must do whatever it orders, and patiently submit to any punishment that it imposes, whether it be flogging or imprisonment... *And that if it leads you out to war, to be wounded or killed, you must comply.* (Plato, 1954: 91. My italics).

To what can we attribute the power of such a text to affect Doukas in this profound way? It would be too easy to argue that he simply read the text within a tradition of conservative political thinking in which the state reigns supreme in matters of individual duty, and, having read the text 'correctly,' drew the political conclusions lodged within it. Socratic philosophers have been interpreted in ambivalent and contradictory ways by both conservative and liberal thinkers (Kraut, 1984: 11-12; Wood & Wood, 1978: 258-261). Modern Greek education placed such texts on an ideological pedestal. These were texts of unquestioned wisdom and moral precept, all the more powerful when read in the ancient language, the ultimate in sanctioned "distinction" in 19th century Greek education (Bourdieu, 1971: 200; Argiropoulou, 1995: 42-44). In this sense, the example of Socrates does not exert its power over Alekos simply through the efficacy of its argument but as a sanctioned model and source of authority. Crucially important are not so much the actual arguments for duty and obedience but how Alekos in this moment of history interprets the model.[26] Following de Certeau's (1984: xii-xiii) theorisation, Doukas' decision to enlist can be seen as an act of cultural consumption, an act of "*poiesis*", and not an over-determined event in which the subject passively reproduces the 'truth' of a totalising discursive system. Having no arguments with which to counter the call to duty, Doukas was in the grip of a crisis of conscience in which arguments of obligation to family and country, delivered by a powerful model of education, prevailed.

Doukas' thirty months of combat experience at the front and the two bullet wounds he received to the head were transformative. They led to post-traumatic crises with which he struggled for the rest of his life. His experiences illustrate Eric Leed's (1979) analysis of the "liminality" and "disjunctions" of war, and how discursive dichotomies such as war and peace, front and home, army and civic society operated in the psyche of

[26] Given the dominating influence of Hegel on Greek philosophical discourse in the 19th century and the former's belief that the state was the "highest of all social institutions" the ideological slide between nation and state becomes an inevitable one (Aiken, 1956: 79-81) See also Argiropoulou (1995: Introduction).

veterans. Leed's hypothesis of World War One as a profound experience of industrialisation through the mechanisation and proletarianisation of soldiers, can be applied to Doukas both in an immediate and long-term sense. Wartime trauma added a further disturbing dimension to the rupture that exile and refugee experience represented. Its effects, deeply embedded in his writing and thinking, continued well after his entry into Australia.

The Crisis Years

After the war, Doukas was hospitalised in Athens for six months, and following a short stay in Mytilini with his family, settled in Xanthi, Northern Greece, where he worked in the family wholesale olive oil business. In this period Doukas tried his hand at farming, which represented more a form of mental therapy and romantic retreat than an occupational vocation. In 1926, after the family business came close to bankruptcy, he took up a clerical position with the Refugee Settlement Commission in Xanthi near Thessaloniki.[27] From 1924 onwards, Doukas tried repeatedly to migrate to Australia, but various factors and events frustrated his desperate urge to leave Greece for the adventure of travel and a new life. His ambivalent rationalisations, emotional states and dreams arising from the desire to migrate, provide an insight into the subjective aspects of the "push" factors of Greek migration in this period.

The years 1922-1927 were undoubtedly a time of mental and physical recuperation for Doukas, with recurrent crises arising from his traumas. His search for philosophical, spiritual and literary ideas was a way of navigating through such crises and providing meaning to his inner turmoil. These searches are manifest in his close correspondence with his brother Stratis, his mentor, confidante and advisor on his mostly unpublished literary output. As a traumatised veteran and refugee, Doukas embodies a complex array of discursive movements. Only a detailed mapping of the cultural and intellectual landscape of Greece in this period, locates his thinking and writing in a comprehensible framework.

The Greek interwar period is often signified by terms such as crisis, military republican rule, capitalist consolidation, refugee settlement, and aesthetic and literary renewal (Chatzeioseph, 2002, 2003; Vergopoulos, 1978; Moullas, 1993). The 1922 military defeat, swiftly dubbed with the supra-human appellative "The Catastrophe", exacerbated the "national schism" which in the interwar years became associated with the division

[27] See Kontogiorgi (2003) and League of Nations (1926) on the role of the Commission.

between refugees and locals, a deep social and cultural divide that at times erupted in violence (Katsapis, 2003; Yiannakopoulos, 1992: 36; Voutira, 1997: 119-120). The rural settlement of refugees in Northern Greece was not simply a temporary measure; it had territorial, economic and ideological motivations. It aimed to Hellenise the new territories, as well as being a social strategy for dealing with the growing influence of communist ideology (Voutira, 1997: 117; Kontogiorgi, 1992; Petmezas, 2003: 204-207). The interwar rural reforms created a large number of struggling small landholders, a situation that generated a certain rural radicalism, an ambiguous mix of socialist and conservative views (Vergopoulos, 1978). Doukas' embryonic political outlook and neo-romantic ideas on nature and social retreat can be largely located in this ideological brew.

It is, however, primarily within the literary discourses of the time that Doukas' thinking can be located. His personal crises led him to a neo-Christian and spiritual eucumenicism that, while connected to broader European movements, also had a distinct Asia Minor refugee intellectual and literary expression. In Doukas' case, this was not a revival of traditional Orthodoxy but a form of modernism and spiritualism that drew on Eastern religious writers such as Rabindranath Tagore. One of the conspicuous influences on him was the Aivali painter and symbolist writer Fotis Kontoglou, who, unlike Doukas, went on to embrace the strict stylistic aesthetics of traditional Orthodoxy in his iconography and writing. Both of these men were deeply influenced by the Greek post-1922 symbolist and neo-romantic movement whose defining characteristic was the aesthetic of escape from mundane social reality.[28] Closely associated with the "generation of the twenties", symbolism contained a strong element of social opposition, a heterogeneous mix of anti-liberalism and anti-classicism, as well as an identification with anti-bourgeois rebelliousness (Dounia, 1996: 29-53). The influence of Scandinavian and Russian literature gave rise to the popular genre of "vagabond fiction" (Chourmouzios, 1940) and proletarian social realist writing (Moullas, 1993: 51-59). These literary movements had a preference for shorter, more interior prose genres that expressed new forms of subjectivity. This is evident in both Alekos and Stratis Doukas' writings in the 1920s, and certain impulses in symbolism undoubtedly served as stepping stones for the ascendancy of the "generation of the 1930s" and the emergence of the urban novel (Moullas,

---

28 Vergopoulos (1978: 165-166) traces the symbolist influence from the beginnings of the 20th century and partly explains the rise of the movement in the 1920s to the similarities between this period and France of the last decades of the 19th century. Both were periods of rapid change and consolidation of capitalism, a change from "old-liberal capitalism" to a "newer interventionist State".

1993). Both brothers favoured the genre of the lyrical prose poem, and Alekos composed short narrative texts and a longer prose work "Phrygia"[29] in this vein. This larger work is a mix of vagabond fiction, older models of ancient Greek adventure or romance novels and an autobiographical narrative of religious redemption reflecting Alekos' reading and his amalgamation of traditional and modern genres.[30] The influences of the 1920s are crucial in understanding Doukas' literary efforts in Australia as engagements with new literary environments, and as fusions and transformations of earlier literary aesthetics.

The mapping of the discursive world of educated Greek society in the interwar period is a large project but one that bears fruit in the study of the literary sensibility of a pre-World War Two migrant such as Doukas. Literary movements in this period often overlapped with political discourses. Symbolist literature spanned the field from anti-political escape to an interest in marginal and proletarian sectors of society, and Doukas inhabited an ambivalent and contradictory location in this field throughout the 1920s and 1930s. While the literary and cultural debates in the 1920s are distinctly Greek in their formation and expression, it is important not to view them as a case of historical exceptionalism (Gallant, 1997). Greek intellectual culture of the 1920s cannot be understood without its European and international dimensions, even if, or exactly because, this takes the form of a fierce debate between nativists and Europeanists or internationalists.

The Voyage South

The descriptions in Doukas' letters home of his voyage to Australia in late 1927 constitute a narrative of transnational Mediterranean and Greek diasporic connections on the trans-Indian oceanic route.[31] The letters refer to numerous acquaintances, both from Asia Minor and Greece, connected to or travelling to various diasporic destinations (Egypt, Havana, Tanganyika, South Africa). They testify to the phenomenon of an active Greek diaspora, both "mobilised" and "proletarian", on the move in the 1920s.

---

[29] Composed in 1924. See manuscripts sent in correspondence with his brother. Stratis Doukas Archives. *Op. cit.* Indicative of the terminological indeterminacy of the period, Doukas never specifies its generic identity.

[30] Some of the texts that provide a background to this work are: Knut Hamsun *Pan, Hunger, Victoria*; M. Gorky *Creatures That Once Were Men*; Lucian *Dialogues*; R. Tagore *Gitanjali*; L. Tolstoi *The Kreutzer Sonata*; O. Wilde *The Picture of Dorian Gray*; *Old and New Testaments*; V. Alfieri *Saul.*

[31] Letters dated from 24 September 1927 to 4 November 1927. *Op. cit.*

The voyage sequences in the letters comprise a narrative that belongs to the genre of travel writing within the context of Western exploration and discovery. This narrative is rich with examples of the Orientalist imagination of a subject in thrall of the European colonial project. This is clear in Doukas' enthusiastic description of the canal town of Port Said and the statue of the French engineer Ferdinand de Lesseps who "[at] the canal's entrance... stands tall on his pedestal pointing in a manly way to the world, the road to the south".[32] As an example of Orientalist discourse at work, a Greek teacher from Port Said had written only a year earlier about "the huge bronze statue of de Lessep which grandly points to the seafarers, the strait that separates two continents and joins the two seas" (Papadopoulos, 1927: 343), thus illustrating Said's (1978: 95) point about the symbolic power of Orientalism that can generate a figure who literally cuts through the Orient.

It is important, however, not to move too quickly over the apparent anomaly of a penniless Southern European migrant in the 1920s identifying with the white man's 'civilising' mission. It is here that the study of so many converging discourses can offer us valuable interpretative tools. The connections are multiple, as we have seen: the influence of Western conceptions in Greek national ideology; the foundational myths, entrepreneurial activities and educational ideologies of Greek communities in the Ottoman Empire; the refugee settlement and Hellenisation of northern Greece; the Orientalising discourses of Western expansion, and in particular, the "quest or voyage motif" in Western literary tradition (Said, 1993: 253-254).

Further, Doukas' epistolary narrative contains distinct elements of an exodus and expulsion myth so central to the foundational narrative of white settlement in Australia (Curthoys, 1999). In heroic vein, Doukas describes the tribulations of Magellan, Cook and Tasman, those "great" European seafarers who were "struggling to find new lands and asylum for fortune's outcasts".[33] Of interest here is that the "story of the Fall and expulsion from Eden, and the story of Exodus from Egypt for the promised land" that co-exist in the Australian national narrative, also echo elements from the narrative of the forced departure of Greeks from Asia Minor, although in the latter case it is the idea of expulsion from Eden that dominates (Curthoys, 1999: 5).[34] In subsequent letters home, where Doukas describes the land and people he encounters, the two myths appear interlocked.

---

32 Letter to family. 28.9.1927. *Op. cit.*

33 Letter to family. 10.10.1927. *Op. cit.*

34 For further discussion of such myths in terms of Asia Minor Greeks see Mackridge (2003) and Papailias (2005: 248-249).

The symbolist imagery of the 1920s and the "imperial narrative" converge in the letters of the voyage in a feedback loop of perception and predisposition (Brandabur, 2000: 322). Doukas' travel narrative, like the white settlers' reading of the Australian landscape and the Indigenous people, is not primarily governed by the actual contingencies of contact but has prior roots in the European discourses of colonialist expansion and voyaging.[35] The diasporic subject travels on routes marked by imaginative texts internalised through readings and education. His travel narrative reveals a sense of the 'discovery' of a world already known, but until then only imaginatively possessed.

One of the aims of this chapter has been to trace Doukas' discursive history as an explanatory strategy for the existence of elements in his early writings of European colonialist and Orientalist tropes and narrative modes. It also anticipates the surprising ease with which Doukas accords with so much of the foundational mythology of colonial British Australia in his early years as a migrant. Importantly however, it resists analysing his literary and epistolary writing in the framework of a passive subordinate subject within the regime of White Australia.[36] Doukas' last letter home from his voyage across the Indian Ocean is an example, on a biographical textual level, of the singular experience and discursive baggage of an historical subject as he is being processed by immigration authorities in the Port of Fremantle on 4 November 1927:

> Then the passport *control* begins. British order. The customs officials take over the smoking-room. We enter two at a time. The attendant shows you where to proceed (in French and Greek). When the *controleur* sees in the reference section of my statutory declaration the words "Macedonian colonisation" he turns with a smile and says, "Do you speak English, sir?". "Yes, sir". "Have you been to America?". "No, sir. But I have been in the English service at Salonica". "Very well, sir". The *conversation* ended there and I ran off with my companions into smiling Fremantle.... What can I say, boys? The omens indicate that the best thing I ever did was to come to this beautiful country. I'll battle superhumanly for a year or two. But I'll live the life I've dreamt about since I was eighteen.[37]

The text, located at the very moment of entry of an Eastern Mediterranean subject into White Australia in 1927, is rich in transnational connections. It is revealingly silent about important aspects of this passage of entry which was no mere formality. The health examination,

35 For examples of similar themes in Australian writing see Curthoys (1999).

36 For a critique of the assumption that Southern Europeans arrived in Australia without their own distinctive discursive history (migrant subjects as "*tabula rasa*") see Pugliese (2007).

37 Letter to family. 11.10.1927. *Op. cit.*

ostensibly to check physical health, was also an examination of racial presentation. Persons considered not to be white, or not white enough, were excluded through the administration of the "dictation test" in "a language likely to be effective under the Immigration Act" (Gilchrist, 1997: 237).[38] In Perth, a multi-lingual Greek from Egypt, Peter Michaelides, was for years employed to "meet immigrant ships arriving at Fremantle port and, whenever necessary, to administer the language dictation test" (Kanarakis, 1997: 41).[39] This racial screening appears not to register in the biographical narrative. There is no disjunction between the subject's European sensibility and the process of passage into white society.

The text, on the contrary, reveals the subject's expectation of participation in a colonial project, the positive attitude to "British order", the association of refugee settlement in Northern Greece with Australian colonial settlement, and a migrant and refugee dream of entering a paradise where past traumas might disappear and dreams be fulfilled. In short, it reveals a subject history *before* the discursive and institutional construction that such a subject is about to undergo in Australian society.

Epilogue

Alekos Doukas' story is partly about how prior expectations and influences play out in his subsequent life but also how the experience and thinking of a specific immigrant subject defies the stereotypical. It is about how the rich discursive history embodied in such subjects engages in a complex and often contradictory way with new discourses and power regimes. Doukas' later engagement with the Australian communist literary movement adds a further level of complexity to his life story in that his fictional works write the past according to another vision that idealises racialised social relations. The literal reading of his fiction as documentary testimony fails to attend to the mythologising functions of such a narrative. Close attention, however, to Doukas' private and public texts and their discursive affiliations leads to a reconstruction of a life that is full of contradictions and multiplicity. The transnational and cross-cultural approach I have sketched runs counter to the monocultural and subordinated representation of the Southern European immigrant in mainstream history. But importantly it also runs counter to a tendency in

---

38 Gilchrist (1999: 235-238) reports cases of Greek refugees from Kastellorizo between 1917-1918 whom the government wanted to refuse entry by applying the "dictation test".

39 Michaelides worked part time in this capacity from 1903-1924 (Kanarakis 1997: 41). Kanarakis includes Michaelides in his nine portraits of outstanding Greek migrants in Australia and the dictation test is mentioned without comment, however, on its significance in the policing of "White Australia".

ethnic history writing to search for univocal historical figures to fit a neat, and often heroic, narrative of a migrant past.

Imagining Italians Abroad

The 2008 Italian Political Election Campaign in Australia

Ilaria VANNI

Senior Lecturer, PhD, at the University of Technology, Sydney

Neighbourhood

The Five Dock Retired Servicemen League (RSL) Community Club is an imposing building on the main road cutting through Inner West Sydney. The Five Dock RSL looks like many clubs in Australia; it is a darkened space filled with the ding and reddish lights of poker machines. There are wide air-conditioned staircases, bars and halls used for shows and for community events, a cafe restaurant that offers a selection of Australian and Chinese meals, and a children's playroom.

What distinguishes Five Dock is that it is considered by many the real heart of the Italian community in Sydney (RAI International, 29 March 2009). This particular RSL happens to be in an Italian neighbourhood; the club sits opposite a row of shops: a *pasticceria che sembra proprio di stare a Roma, il negozio di deli Ranieri, il fruit market*. Of the population in the surrounding neighbourhood, 31% were born outside Australia and 30% speak Italian, Greek, Cantonese, Mandarin, Korean or Arabic (Five Dock RSL, 2009: 6).

But on the evening of 26 March 2008, the conversations at the Five Dock RSL Community Club were 100% Italian. The nondescript darkness of the upstairs hall was broken with splashes of Italian colours: flags, brochures artfully arranged in fans, badges, business cards with Italian names. Italian men and women – largely of the first generation of Italians to arrive in Australia after Second World War – stood in small groups, dressed in trousers and short sleeves shirts, chatting in assorted accents, every now and then dropping the tone down to a sottovoce that signaled the exchange of some juicy gossip. In contrast, men wearing suits and a few women – still first generation immigrants to Australia, but much later arrivals – did the rounds among the groups, as if they were the bride's parents at a wedding.

The gathering at Five Dock RSL was the official launch of the campaign for the Partito Democratico (PD, the newly formed centre-left Democratic Party), which was running candidates for the Italian political elections a month after this event. The candidate for the Lower House, Marco Fedi, and for the Upper House, Nino Randazzo, were to present the electoral program of the PD for the *Italiani all'estero*.

Much happened before and after the launch itself. The newly constituted *direttivo del partito* performed their own deep play along the shifting lines of power among the various diverse political constituencies within the spectrum of the Italian institutional Left. *I compagni della FILEF* (*Federazione Italiana Lavoratori Migranti e Famiglie*), who much contributed from the 1970s to the making of an Italian diasporic political community, sat discussing the defection of another *compagno della FILEF* to the list of *Rifondazione Comunista* (which in the Italian political spectrum situated itself to the left of PD). On the right hand side others more or less connected to the new formation of the PD, and some who as it appeared clear when the floor was declared open for questions – aligned with the Right party *Popolo della Libertà* (PDL, the new name of Berlusconi's Forza Italia). The candidates, both wearing smart casual, sat a few metres in front of the audience, behind a table.

The presentation of the party's electoral platform was smooth, conducted with professionalism. The themes covered included telecommunications, double citizenship, retirees' rights, and the promotion of Italian culture and language. At question time, those aligned with the *Popolo della Libertà* provoked the candidates with an array of personal questions revolving around the money spent for the electoral campaign, as well with some queries about the party's platform itself. *I compagni della FILEF* – arms crossed – replied booing and mumbling loudly. Marco Fedi, who at the time had served almost two years in the Italian Parliament and was running for re-election, replied calmly, explaining what had been done, what had not been done and why, and what needed to be done. After the presentation, a few people remained. Hands were shaken. Then the candidates, the executive of the Sydney branch of the PD (a voluntary executive) and some *compagni della FILEF* strolled to a local pizzeria for dinner. On the street, Marco Fedi stopped for another handshake, and another calm explanation.

On the same street three days later – mid-morning on Saturday 29 March – the delis were crowded with Italian women doing their weekend shopping. A blonde woman with a blue top, white jacket, pearls and sunglasses and a younger woman in jeans and blue tank top stopped *le comari e i compari* on their way to the next shop, distributing leaflets. The older woman saw a car pulling up, and took the opportunity to open the door and help another woman out, before giving her a leaflet that she promptly stuffed into her handbag.

A taxi pulled up, and out of it stepped Luigi Casagrande, the PDL candidate to the Senate, dressed in khakis. He was met by a television camera and an interviewer from RAI International – the Italian Government cable and satellite channel broadcasting to Italians living abroad, and broadcasting news of *Italiani all'estero* back to Italy. RAI's transformation from cable and satellite to free-to-air TV was at top of the electoral platforms of both parties.

The camera zoomed in on the blonde woman, Teresa Todaro Restifa, the *Popolo della Libertà* candidate for the Lower House. She explained that Italians in Australia should vote for her party and in particular for herself, as she will maintain promises, like that of the right to double citizenship, since, she noted cheerfully, "It is what our countrymen most want – they want to die Italian".

Casagrande reiterated to the interviewer that the platform was not important – what was important was that results were delivered. This would be his refrain throughout the campaign (in another interview, when asked to name his party, he could not answer, Teleglobalaustralia, 2008).

The camera zoomed out on the group, sitting down at a table drinking coffee. The interviewer told viewers that the candidates travelled as far as South Africa to campaign and that a great friendship and reciprocal esteem has developed between the South African candidate, second on the list. Then the group was off to Melbourne, to meet another community.

The dimension of these meetings remained bounded by specific localities, intended following Arjun Appadurai as "relational and contextual rather than as scalar or spatial... a complex phenomenological quality, constituted by a series of links between the sense of social immediacy, the technologies of interactivity, and the relativity of contexts" (Appadurai, 1996: 178). In this sense, locality includes virtual as well as physical communities, and it is complemented by the idea of neighbourhood as "the actual existing social forms in which locality, as dimension or value, is variably realised. Neighborhoods, in this usage, are situated communities characterised by their actuality, whether spatial of virtual, and their potential for social reproduction" (1996: 178-179). This is the unfolding in space of the Italian electoral campaign in 2008: a succession of neighbourhoods.

The Foreign Constituency: Italians in the World Vote

> From 26 March we will vote for the election of Camera dei Deputati and Senato della Repubblica. Italian citizens living abroad and registered in the Register of Birth, Marriage and Death of Italians living abroad, will vote by mail in their country of residency. Each Italian citizen will receive the necessary documentation to vote directly at home, from the nearest Consulate.
>
> (From the Partito Democratico in Australia blog, http://pdaustralia.blogspot.com/2008/02/partito-democratico-in-australia.html)

> *L'Italia può riconquistare una posizione di eccellenza nell'economia globale se utilizza pienamente una risorsa troppo a lungo trascurata: gli italiani all'estero.*
>
> (Italy can recapture a position of excellence in the global economy by fully utilizing a long-neglected resource: Italians abroad) (PD/Italians abroad leaflet, political campaign 2008).

The two narratives above capture two moments of the 2008 Italian election campaign in Australia. In early 2008, for the second time in only two years, Italians living abroad or *Italiani nel mondo* (Italians in the world) as they are known in Italy (with the definition invented during Fascism to put a positive spin on migration), took part in the Italian national political elections voting their own representatives to the "Foreign Constituency". They elected eighteen representatives: twelve in the Lower House and six in the Senate. The candidates were chosen among Italians living abroad, not among Italian politicians.

The novelty of this far-reaching form of representative democracy has captured the attention of many political scientists and historians, who have produced detailed accounts of the mechanisms and historical process behind it (Mascitelli and Battiston, 2008; Bauböck, 2007, 2005, 2003; Biscaro, 2007; Colucci, 2002). While offering an overview of the scheme and its historical background, this chapter is mainly concerned with the campaign that preceded the event itself. The fieldwork was conducted in Sydney during the 2008 brief election campaign. The focus will be therefore on Australia, the country with the largest number of Italian citizenship holders in the Asia-Pacific, Middle East and South Africa electorate district (known affectionately as "*la circoscrizione mostro*", that is, "the monster district" or "the rest of the world district").

The creation of an offshore electorate and the exercise of external voting rights are the result of a century-long process. Over almost one hundred years, diaspora voting rights intermittently resurfaced on the Parliamentary agenda, only to be continuously postponed. First flagged in 1908 during the inaugural Conference of Italians Abroad in Rome, which addressed the adverse conditions a great number of Italian migrants had to face (Colucci, 2002), external voting rights were again discussed in the late 1910s under the Giolitti government, during the early Fascist regime, the mid-1950s after emigration from Italy had resumed, and again in the mid-1970s and early 1980s at a time when push factors had exhausted (Battiston and Mascitelli, 2008; Colucci, 2002). On each occasion a combination of idealism and opportunistic factors drove the agenda.

Fascist Italy dreamt of using emigration as a means for colonial and commercial penetration abroad (Tintori, 2006). The ruling Christian Democrat party's activism towards migrant external voting rights between 1976 and 1982, when numerous bills were presented in Parliament, was driven by the desire to gain moderate votes as communists were gaining terrain in the national political arena (Colucci, 2002). However, only in the late 1990s was the process to provide Italian citizens living abroad with the constitutionally enshrined right to vote finalised, through the creation of an offshore constituency, subdivided in four global districts: Europe; Southern America; North and Central America; and Africa, Asia, Oceania and Antarctica. Not surprisingly, the latter is conversationally known as "*la circoscrizione mostro*". The implementation of this electoral system required the amendment of a series of Constitutional articles (48, 56 and 57), and approval of an implementation law No. 459 of December 2001.

What changes occurred in the last decades of 20th century to bring diaspora voting rights back in the Parliamentary agenda and more importantly to be finalised? Why was this far-reaching scheme of representative democracy approved almost one hundred years after extraterritorial voting rights were first flagged, at a time when Italian emigration has become a marginal phenomenon?

It is suggested here that a number of factors and elements came into play, including a more globalised world and the emergence of new world order, driven by a neoliberal agenda. In this new global context the concept of migrants – and consequently of difference – as resource finds a new interpretation. According to Hardt and Negri (2000, 198-201) with the emergence of the new, decentralized, distributed world order they call Empire the idea of difference is redefined in terms of inclusivity and managerial control. Difference is first absorbed into the "powerful vortex" of Empire, then imagined as cultural and celebrated, and finally managed and "arranged in an effective apparatus of command" (Hardt and Negri, 2000: 200). In economic terms, with the erasure of nation states' borders, difference in the world market becomes a niche market, a new product, an opportunity:

> With the decline of national boundaries, the world market is liberated from the kind of binary divisions that nation-states had imposed, and in this new free space a myriad of differences appears. These differences of course do not play freely across a smooth global space, but rather are regimented in global networks of power consisting of highly differentiated and mobile structures (Hardt and Negri, 2000: 151).

Translated into Australian terms this myriad of differences produces a gourmet multiculturalism, a multiculturalism to be enjoyed, consumed and celebrated. Australian anthropologist Ghassan Hage (2003) has stressed the aesthetic and fundamentally middle class character of this type of multiculturalism. Rather than promoting migrants rights this institutional multiculturalism is at the service of the economy, driven by the constant search of new market niches and productive opportunities. Diversity is something to be managed: a productive resource able to provide new market opportunities (Hage, 1998, 2003; Zizek, 1997).

The same logic is reversed and applied to the Italian diaspora, which is seen as a likely source of economic growth for Italy. Mirroring the neoliberal discourse of migrants and multiculturalism as an economic resource for the host country, diaspora Italians are seen as potential commercial networks, partners and markets which can boost Italy's trade and international position. These are, for instance, key points in the manifesto "*Azzurri nel mondo*" (Azzurri in the world) produced by *Forza Italia*, the previous name of PDL (Forza Italia, 2003). "*Moltiplicare la forza dell'Italia*", "*Gli italiani nel mondo come risorsa economica, culturale e politica*", "*L'esportazione delle associazioni di categoria*", "*Il rilancio delle imprese italiane all'estero*", "*L'osservatorio sulla valorizzazione della presenza italiana*" are some of the subtitles in the document (see also Biscaro 2007).

The conceptualization of diaspora Italians as a resource to manage was shared across the political spectrum, as exemplified in the PD leaflet quoted at the beginning ("Italy can recapture a position of excellence in the global economy by fully utilizing a long-neglected resource: Italians abroad", PD/Italians abroad leaflet, political campaign 2008), and it is because of this shared vision that the Italian Parliament legislated in favour of migrants' voting rights.

Campaigning and Imagining

By focusing on the Australia-based Italian political campaign, the discourses of the candidates, the interaction between candidates and potential voters, it is possible to analyze the interplay of the above mentioned elements into migrants' everyday practices. Following the campaign as it unfolded in public events and was captured and spread in participatory media through websites, direct email campaigns, blogs and podcasts this chapter examines the making of a diasporic political subjectivity largely shaped through a politics of imaginary (Neilson, 2005).

The deployment and modulation of imaginary, along with its correlations of affects, sentiments, intimacy and fantasy, was used by both parties as the main campaigning modality with subtle differences. In this sense the technique, the style and form of the campaign could be seen as being as relevant to the political project as the contents that were transmitted.

The imaginary that was constructed in meetings, repeated in the digital agora of YouTube and podcasts was of a specific type: a diasporic imaginary, set in an imagined Italian community. Central to it is the idea of being at once both dislocated and located, and of recontextualised temporality that allows Italian citizens in Sydney and their elected representatives to be both in Australia and Italy at the same time.

Benedict Anderson writes in *Imagined Communities* that the spreading of print media gave a form to the sense of "homogenoeous, empty time" that he identifies (with Benjamin, 1992: 252) as the beginning of modernity and of the idea of nation. The novel and the newspaper provided a platform for a shared simultaneity among a community of readers. This community was imagined as the consumption of both novels and newspapers remained largely an individual matter, although it created shared worlds of bounded horizons (Anderson 1991: 23-30). What happens if we move this preposition to an Italian neighbourhood in 2008, and to the shift in political campaigns to participatory media?

If print-capitalism was fundamental for the origin of modern nations, how does high technology capitalism (Dyer-Whiteford, 1999), thriving on flexibility, outsourcing, immaterial labour and diversity, influence the way a diaspora imagines its "nation"? The dislocation of a diasporic community is reterritorialised in unbounded horizons, which bring together at the same time being in Australia and being in Italy through relations, everyday practices, media and now, for the first time, Parliamentary politics. Conversely, a particular discourse of migrants as added value makes possible the reabsorption of *Italiani nel mondo* into the nation.

Both PD and PDL deployed the trope of migrant productivity. "A source of excellence" was a motto often reiterated by PDL candidate Casagrande in his media interviews and meetings with the Italian community. This expression, as insightfully noted by *La Fiamma* commentator James Panichi, suggests with its typical managerial tone an association with an economic/productive activity: excellence in manufacturing (19 March, 2008: 21). An article that appeared in *La Fiamma* reported on the PDL campaign was titled: "To look at reality from an international angle. Italians abroad; a wealth of professionalism, knowledge and talent valuable to the mother country" (Tornari, 20 March, 2008: 29). Restifa went further, proclaiming in her public speech during the official launch of PDL in Australia that it was the 60 million Italians living abroad that created the Made in Italy (28 March, 2008). The PD stressed the importance of productivity at the top of its electoral platform: "Italy can recapture a position of excellence in the global economy by fully utilizing a long-neglected resource: Italians abroad" (Partito Democratico, 2008b). Italian migrants, in this discourse, are not simply a resource for the host country, they become an asset for Italy itself. Their reterritorialisation within the unbounded horizons of Italy is mediated by their ability to act as conduits in the global flows of capital.

If the act of voting for the Italian government might have had little relevance for many at a political level (the percentage of voters in Australia was of 37,002% for the Senate, 35,852% for the Parliament, against the 80,404% and 80,512% respectively in Italy, Ministero dell'Interno, 2008), it provided an important symbolic tool. Simultaneity – which is possibly the counterpart of the much-discussed post-colonial in-betweeness and which has always made possible a "double presence" in everyday practices, imaginary, memories and desires – was legitimised by the seal of approval of official politics. *Gli Italiani nel Mondo*, who until now were conceptually relegated to the waiting room of geography as much as of history, had been symbolically reabsorbed into Italy.

The imaginary at the core of this symbolic mutation is situated in the everyday as "both memory and desire", as Arjun Appadurai writes of diasporic imagination "into mythographies different from the disciplines of myth and ritual of the classic sort (1996: 6). The ability of diasporic imagination to produce figurations that are both in conjuncture and disjunction with the mythographies of the country of origin, are linked by Appadurai to the pervasive influence of mass mediated imaginaries that traverse national spaces (1996: 6). Having recognised the autonomy of diasporic imaginary in its production of memories and desires, what happens when we move it into the plane of institutional politics? What is the relation between the memories and desires exchanged while shopping in an Italian neighbourhood in Sydney and Italian Parliamentary politics? Is this diasporic imaginary simply influenced by a mass media, or is it able in turn to influence media representations?

The mythographies produced in the diaspora in virtue of their ordinariness mobilize a set of representations and self-representations that has very little to do with the grandeur of Italian mainland political imagination. The Italian diasporic political community is imagined in everyday objects and practices: consuming Italian products, watching Italian television, speaking Italian, shopping in Italian stores. These practices speak of an embodied imagination that is entangled with senses, affect and symbols rather than being determined by political ideologies or citizenship practices.

Citizenship in itself becomes here a central question: how it is enacted, and what does it become when it is unhinged from notions of nation/state, common language and culture and belonging to a territory? Citizenship is intended not simply as a bundle of rights and duties, endowments and entitlements – the faithfulness to the Italian Republic that one promises upon becoming a citizen for instance – but as a floating signifier that covers everyday cultural practices, senses of belonging, and participatory politics as well as the more traditional legal and political definitions.

What emerges is a circular movement of citizenship from an engagement with the public sphere to retreat in the private realm. The actual voting mechanism, *direttamente da casa* (directly from home) via postal vote in opposition to the ritual of voting in a public space and under very strict rules, is a telling symbol of this movement.

The PD platform consisted of six points: circular and reciprocal information from and to Italy via a free-to-air RAI International; promotion of Italian language and culture; reacquisition of Italian citizenship; restructuring of the consular networks designed to employ local Italian citizens; tax reform on properties owned in Italy by Italians living abroad; better recognition of the standards of excellence achieved by Italians worldwide, including educational exchanges (*Partito Democratico*, 2008b). The PDL's platform promoted the reacquisition of citizenship; pensions; social welfare; specific programs to facilitate young Italians; information; promotion of Italian culture and of Made in Italy (PDL Australia, 2008).

These platforms were mediated in images, narratives, speeches, discussions that resonated in old media in the Italian newspapers in Australia, *Il Globo* and *La Fiamma*, and in the programming of the radio station associated with them, *Radio Italia* and on the Australian multicultural radio *SBS Italian Program*. They were repeated in participatory media, in blogs, websites, podcasts and YouTube videos. If some of the themes were common to both parties, the way the two parties decided to present their programs and convince the electorate was not.

A short ad to publicize PDL for instance is shot in a non-descriptive Italian neighborhood. A catchy tune plays in the background. A group of three people, one elderly man and woman and a younger woman sit at a table in a cafe eating *paste* and drinking espresso. The voiceover invites the electorate to choose PDL for "concrete results in important themes" (*risultati concreti per i grandi temi*). Keywords from the PDL program run at the bottom of the screen. Cut to an indoor scene of the same cafe, where a young couple sits on a stuffed sofa cuddling their baby next to an Italian flag and several objects. In the next room a group of young people sits at minimalist tables eating gelato, chatting and looking at passports. Cut to a blue background to the lower house and upper house candidates – Casagrande wearing casual clothes, Restifa in pearls and a blazer of the same sky blue of the party's logo – invite the electorate to vote for PDL (PDL Australia, 2008).

The PD's electoral videos, one for each candidate, in stark contrast, are shot as a close ups against a neutral background and next to a folded PD flag, which being white, red and green is immediately associated with the Italian flag, placing the ad in an official and institutional space (Fedi 2008; Randazzo, 2008). In one of the two videos, Marco Fedi looks directly at the camera, and consequently at the viewer, blinking a little, and speaking in a calm and steady tone. He reminds the electorate that it is possible to change Italy by voting PD, a new party that will give governmental stability to the country and to Italians living abroad. "The first message that I want to send you all", he says, "is to vote. Do not give in to desperation, there is hope and the hope is Partito Democratico" (Fedi 2008). The video continues explaining how change can be built together by Italians in Italy and those living abroad, and it illustrates what he has achieved "with honesty and integrity" as a Member of Parliament for the Foreign Constituency in the government that just dissolved.

PDL chose to represent *gli italiani nel mondo* by deploying an array of characters that, taken together, stress the continuity of Italian families and traditions (elderly/young couple) together with the image of the future (a group of young people). Their Italianness is codified through a very ordinary practice – so ordinary it is a stereotype. They sit in groups in a cafe, eating Italian food. This ordinariness contrasts with the official content of the voiceover and the titles running at the bottom of the screen. It is also interrupted by a scene towards the end of the ad, where a young girl opens her Italian passport.

In contrast, the PD ad of Marco Fedi's electoral speech (as well as that of Randazzo, since the videos are shot in the same location and in the same format) is instantly recognizable as a campaign video, both through the institutional signs – the flag and the superimposed logo of the party – and through Fedi's performance. By looking directly into the viewer/elector's eyes Fedi erases the distance between himself and his voters, establishing a personal relationship emphasised by his invite. Intimacy is reinforced by his current blog, and by the emails, which he sends regularly and directly to his electorate with updates from the Italian Parliament. The content of these emails is also published as a column in the newspaper *Il Globo* (Fedi, 2009). The same erasure of distance between the politician and the electorate was also enacted at the official launch of the program narrated earlier. In his electoral video, Fedi appears "as he is", the camera angle and close up revealing his face against a minimal, if not bare, background.

Ann Gibbs, in her study of the Australian politician Pauline Hanson, has described the use of close ups as opening up a space of "affective contagion" where the viewer find herself or himself catching on the feelings and emotions of the speaker, and at times repeating her or his involuntary movements, such as blinking, staring, gesturing. According to Gibbs, in fact, it is through the face that we communicate our affective state to others and to ourselves via feedback (Gibbs, 2002). Fedi, in this sense, communicates through his face and tone a range of sentiments that mobilize an imaginary of honesty, integrity, dignity, trust, intimacy. It is exactly through the modulation of these affects within a political project that it becomes possible for an imagined community – which does not have a shared, bounded territory because of its own dislocation – to relocate itself in a virtual proximity to Italy via the relationship with its elected representatives.

The affective and relational aspect of voting over official political programs and parties is summarised in a comment on the newspaper *La Fiamma*:

> Like many others I am asking myself who would better represent me (represent us) at a political level, and I realised that there are many who just want to reach a seat in Parliament or in the Senate in Italy. Each of them has some form of credibility… No choice will be based on the party they represent, nor on wealth. I put in the first place a proved human motivation (Pollicina, 2008).

The erasure of distance via means of mediated affects is also central in establishing a live link between Italy and *Italiani nel mondo*. The insistence of both parties in promoting a free-to-air RAI International is not accidental, as the channel ensures a flow of information from and to Italy. The language used to describe belonging to the same community is an affective one. Fedi stresses the importance of collaborations between *Italiani in Italia* and *Italiani nel mondo* in the political project (Fedi, 2008). Luigi Casagrande, in an interview appealed to *vergogna* (shame), a very powerful sentiment in the Italian imaginary, to describe what "we *italiani nel mondo*" felt about a rubbish emergency in Naples brought about by a combination of lack of garbage processing plants, *camorra*, and strikes. In the same interview he declared that it is only through the unity of Italians in Italy and those living abroad that Italy is a winning country (Teleblogaustralia, 2008). Nino Randazzo pleaded for support and backing for the good of Italy and Italians in the world (Randazzo, 2008). Restifa, as noted above, declared in an interview to RAI International that it is important to re-acquire Italian citizenship because what Italians most want is "to die Italian" (RAI International, 2008).

It is this constant shift from the enunciation of the official political platform, and a politics of imaginary centred on intimacy (either lived in community meetings or mediated) that created the possibility of a diasporic political community.

Conclusion

This essay has considered the campaign for the Italian political elections in 2008 as an example of a politics of imaginary built around everyday practices and affects, articulated and enacted through staged and non staged meetings, launches, official speeches, programs and the use of a range of diverse media. I argued here that the peculiarity of a diasporic community electing its own representatives resides not simply in the political and legislative mechanism, but also in the everyday workings of the electoral campaign. The result is, as we have seen, the production of a politics of intimacy and affect, enacted both by the use of local and participatory media and by the tone, tempo and texture of the campaign itself, often in contrast with the grandeur of the discourse that oozes out of the official platforms of all parties and in direct opposition to the size of the electorate.

La circoscrizione mostro brings together very different geographical and socio-cultural areas, spanning from the Pacific to Africa. Territory plays a major role in the campaign and the modulation of distance reflects the tension between the public sphere and private aspirations, the use of participatory media and its rhetoric as a tool to reach beyond the borders and across networks, Italy as a nation and *Italiani nel mondo* as an imagined and imaginary community.

Transmediterranean Dispersals

Mazzini, Hindu Nationalism and Sonia Gandhi

Goldie OSURI

Lecturer, PhD, at Macquarie University, Sydney

Introduction

When I was invited by Joseph Pugliese to contribute to the symposium, "Relational Dis/Locations: Mediterranean Cultures in Translocal and Transnational Contexts" which forms the basis of this edited collection, I was not certain that I would have much to contribute. I don't consider myself a scholar working in the area of transnational Mediterranean cultures or diasporas. Memories of my mother proudly stating that Telegu (the language of the state of Andhra Pradesh where my family comes from) is the Italian of the East vaguely flitted through my brain. I wasn't sure I would be able to present anything substantial on that issue, apart from a postcolonial line of questioning why Italian couldn't be called the Telegu of the West. I vaguely flirted with the notion that I could engage in an act of "provincialising" Italian by way of such a critique. But as I thought a little more seriously about the invitation, I realized that it offered an opportunity to discuss an aspect of a current project that I have been working on – the politics of Hindu nationalism as it concerns its anti-Christian, anti-conversion campaigns. In relation to the symposium, it occurred to me that there was a transmediterranean connection, disavowed by proponents of Hindu nationalism, which needed to be discussed in relation to Hindu nationalist representations of another transmediterranean figure – that of Sonia Maino Gandhi – leader of the Congress Party and a key player in India's political landscape. Furthermore, Gandhi's own engagement with those representations, her self-styled embodiment as widow, mother and daughter-in-law of the nation conforming to a Hinduised politics was worth discussing in this context. The symposium's invitation to examine "how Mediterranean cultures are engaged in flows of transnational cultural exchange" and "are reconstituting local cultures and identities and translocating these things in the context of national/ist cultures" seemed perfect for this occasion (Seltin, 2009). Unfortunately, I was unable to participate in the symposium, but I thank Joseph Pugliese for the opportunity to present this essay on transmediterranean connections, Hindu nationalism, and Sonia Gandhi's incarnation as daughter-in-law of the nation.

What does the attempt to map a transnational cultural exchange, a circulation of an idea, or the transnational movement of a person offer? In other words, what is the point of looking at those transmediterranean connections to Hindu nationalism? At the very least, it unsettles the contiguity between geographical territory, nationalist claims, and a racialised embodiment. But beyond this unsettling, it offers the opportunity to illustrate the promiscuity of cultural forms through what Benjamin Lee and Edward Lipuma (2003) call "cultures of circulation" and the manner in which these cultures contest fictions of purity and homology – of blood, body, territory and nation. These fictions of purity and homology are often used not only to exclude otherised identities within a given nation-state, but to enact and justify violence against those bodies – often racialised, religious, gendered or sexual others constructed as anti-national. In this sense, it is extremely important to map the promiscuity of cultural practices and ideas as they offer a way to contest arguments and justifications for violent otherising practices.

The concept of "transnational connectivities", theorized by Inderpal Grewal, enables this mapping. In her recent book, *Transnational America: Feminisms, Diasporas, Neoliberalisms* (2005), Grewal states that the term "connectivities" recalls "collectivities" and can illustrate how "transnational connections" can "produce groups, identities, nationalisms" (2005: 23). Grewal's theorisation of this term serves an alternate way of thinking about the "global" (2005: 22). This alternate conceptualisation is a necessary one, she argues, for while the "will to globalization... was both profoundly cosmopolitan as well as imperialist... 'global' capitalism did not constitute the totality of economic or social relations that were existent or possible" (Grewal, 2005: 22). The term "transnational" which is "more humble, and often a more adequate label for phenomena which can be of variable scale and distribution", therefore, appears to describe specificities of connections across geographical or cultural spaces (Grewal, 2005: 22). While contemporary connections across national spaces may be termed transnational, it is also necessary to reference how transnational connectivities are often built on "older histories of colonialism and modernity" (Grewal, 2005: 24). These older histories – an earlier will to colonial expansion that precedes the will to globalisation – form foundational structures for examining many contemporary formations of nationalism. In fact, the transnational Italian/Indian connection in the politics of Hindu nationalism that I want to examine is tied to this older history of colonialism and modernity as I illustrate in this essay. Two moments of transmediterranean dispersal offer a way to unpack the fictions of Hindu nationalism's homology between religion and nation. The earlier moment occurs in the influence of the Italian nationalist, Giuseppe Mazzini (1805-1872) on anti-colonial Hindu nationalism. The second moment of dispersal occurs through the localisation of Mazzini's writings by contemporary forms of Hindu nationalism deployed against the political leadership of Sonia Gandhi. In order to chart these moments of dispersal, I will begin by contextualising Hindu nationalism and trace the manner in which Mazzini's writings are localised within an anti-colonial moment. The second section addresses how Indian womanhood plays a key role in Hindu anti-colonial nationalism. This section forms the basis for discussing the second transmediterranean dispersal which I map through a reading of Hindu nationalist representations of Sonia Gandhi, and her embodiment of Indian womanhood.

Transmediterranean Dispersal I: Mazzini and Hindu Nationalism

As Kamat and Matthews describe it, the 'formal origins of the Hindutva movement can be traced to 1925 when the Rashtriya Swayamsevak Sangh (RSS, literally National Volunteer Corps) was founded for "propagating Hindu culture" (2003: 8). M.S. Gowalkar was the supreme leader of this organization. One of its intellectual inspirations was Vinayak Savarkar who wrote about the necessity of the alignment between masculinity, race, culture and nationalism which had to be asserted against perceived otherness or foreignness. Kamat and Matthews cite Gowalkar's argument that "foreign races must adopt the Hindu culture and language, must learn to respect and hold in reverence Hindu religion, must entertain no ideas but those of the glorification of Hindu race and culture" (quoted in Kamat and Matthews, 2003: 9). Vinayak Savarkar argued that the "conceptualisation of Indian national identity must, at its foundation, be based within the political philosophy of Hindutva" (quoted in Chaturvedi, 2003: 169). This philosophy, Chaturvedi emphasises, is not identical to Hinduism. Hinduism is only "a fraction, part of Hindutva", whereas "Hindutva is not a word but a history" (2003: 169). Loosely translated, Hindutva means Hindu-ness, yet its meaning is not confined to Hinduism, rather it seeks to swallow the history of the nation-state and regurgitate it as the identity of a Hindu nation-state. In this sense, Ashish Nandy describes Hindu nationalism as a "modernist creed which seeks to retool, on behalf of the global nation-state system, Hinduism into a national ideology and Hindus into a 'proper' nationality" (1998: 283). While this ideology emerged in the context of anti-colonialism, the reference to "foreign races" and the attempt to subordinate them to Hindu nationalism, as Kamat and Matthews point out, was not only "to the British against whom Indians raged a fierce anti-colonial struggle", instead "it stood above all (and continues to stand) for Muslims in the subcontinent, followed by the entire range of religious and cultural minorities such as Indian Christians, Zoroastrians, Jews, and Sikhs" (Kamat and Matthews, 2003: 9). While Hindu nationalism was largely sidelined in Indian politics until the 1990s, it emerged as a vital force through the electoral success (1998 to 2004) of the Bharatiya Janata Party (BJP) who were allied with Hindutva organizations such as the RSS, the Vishwa Hindu Parishad (VHP-cultural wing) and the Bajrang Dal (militant wing). Despite the loss of Central government power in 2004, the BJP remains in power in some state governments such as Gujarat, Karnataka, and Rajasthan. And, Hindu nationalism is very much a force to be reckoned with in the Indian context. In fact, during the 15th general elections held in India, they led the opposition to the UPA (United Provisional Alliance) led by Congress through their coalition, the National Democratic Alliance (NDA).

Vinayak Savarkar, the anti-colonial nationalist who formulated the philosophy of Hindutva was influenced by the writings of Giuseppe Mazzini. Having participated in anti-colonial revolutionary activities at a young age, Savarkar arrived in London in 1906 for an "advanced law degree" (Chaturvedi, 2003: 161). In 1907, he published the Marathi translation of Mazzini's autobiography. Chaturvedi states that the Abhinava Bharat Society (ABS) which Savarkar participated in may have been modelled after Mazzini's Young Italy, a youth organization for freeing different parts of Italy held under foreign rule using methods of violence and insurrection (2003: 161). Perhaps the most important element of Mazzini's influence in the context of Hindu nationalism was the use of religion as a resource. In *Mazzini: A Life for the Religion of Politics*, Roland Sarti states that for Mazzini "religion was an instrument of political struggle" (1997: 4). Not only was it an instrument, Mazzini recognised it as "a necessary element of revolution, and faith in God the necessary underpinning of political conviction" (Sarti, 1997: 4). In making this link between religion and nation, Mazzini was different from other nationalists. Mazzini's republicanism was neither "liberal nor socialist" (Sarti, 1997: 3). His concern for the unification of Italy meant that some moral force had to bind a nation's citizenry. While Mazzini "rejected the theology of the Roman Catholic Church and its conservative politics", he believed that "there could be no progress toward a just society without the stimulus of religious enthusiasm" (Sarti, 1997: 4). In this sense, Sarti suggests that "insistence on the unity of religion and politics separates Mazzini from the political mainstream of his and our time. He would have agreed with the criticism of religious fundamentalists that modern civilization shows a dangerous tendency to ignore moral concerns" (1997: 4). Mazzini's views would have been inspirational for young Savarkar, not least because upper-caste religious identities in the Indian context were being challenged on a number of fronts.

Upper-caste Hindu panic about the threat to its privileges appears to be due to a number of social changes in the late 19th century. Sumit Sarkar has identified the manner in which early British census reports and their attempts to enumerate religious communities and fix caste rankings had some significant effects. One effect of the census had to do with constructing and distinguishing discrete Muslim and Hindu communities in terms of demographic numbers. Another effect was "a quantum leap in the number of caste associations and of publications by or on behalf of lower caste groups, seeking in the main upward mobility of 'Sanskritizing kind,' but also going in at times for quite a lot of anti-Brahminical and anti-caste rhetoric" (1999: 91). Combined with these assertions, the late 19th and early 20th century also witnessed what has been described in missionary literature as a period of mass conversions where families, villages or even sub-castes converted due to repeated famines (Forrester 1991). Also, a few caste/class rebellions were sometimes aided by missionaries as Forrester (1991) and Sarkar (1999) note. These changes caused a great deal of anxiety "best articulated by U.N. Mukherji's text, *Hindus: A Dying Race* (Calcutta 1909), which skillfully used some census data and predictions to develop a horrific vision of Hindu decline" (Sarkar, 1999: 91). Swami Shraddhananda's later text, *Hindu Sangathan: The Saviour of the Dying Race*, made the issue of conversions central, and justified *shuddhi* or reconversion movements (Sarkar, 1991: 91). Hindu nationalist organizations such as the Arya Samaj, the Brahmo Samaj, the Hindu Mahasabha and the Hindu Sangathan emerged within this context. There were important differences between these organizations as Jaffrelot (1999) indicates. For example, the Arya and Brahmo Samajists attempted to reform Hinduism as a response to the representation of Christianity as a "rational" religion, in the process reworking Hinduism as a rational religion (van der Veer, 2001; Jaffrelot, 1999). In comparison to the reformatory zeal of Brahmo Samajists such as Raja Ram Mohan Roy, the Hindu Sangathan and the Hindu Mahasabha were more overtly interested in maintaining upper-caste privileges. Yet regardless of these differences, a key idea that these movements propagated was the assumption that Hinduism is an originary Indian religion. This idea influenced the emergence of Hindutva ideology, and also remains the mainstay of contemporary mobilisations of Hindutva organisations.

After 27 years of imprisonment for sedition by the British government, Savarkar officially joined the "All-India Hindu Mahasabha and served as its President between 1937 and 1944" – his platform, "Hinduize Politics and Militarize Hindudom" (Chaturvedi, 2003: 162). Hindutva ideology, therefore, not only assumed an originary relationship between religion and nation, religion was to be transformed into a militant resource for a Hindu nation. This militarisation of Hinduism was influenced by a transmediterranean dispersal of the unity of religion and politics. It found fertile ground in Hindutva anti-colonial nationalism which directed its energies not only against a colonial government, but also against those that it perceived to be a threat against uppercaste privilege – those seeking equality such as lower castes, women, and religious others who were rendered anti-national.

Gendered Constitutions of Hindu Nationalism

Like many Hindu nationalists, Savarkar wrote about the role of Hindu women as key to the link between religion and nationalism. Rosalind O'Hanlon discusses the manner in which the domestic sphere served a central role in producing women as devoted wives and enlightened mothers for the Hindu nation (1994). Chaturvedi states that Savarkar contributed to these debates in the essay, "Women's beauty and duty" where he argued that the "primary duty of the woman is to the home, the children, and the nation" (2003: 166). For Savarkar, "any woman who digressed from her duty" was "'declared morally guilty of a breach of trust'" (Chaturvedi, 2003: 166). In fact, as Chaturvedi points out, "public debates on the topic of female improprieties that violated the emergent Hindu norms within the domestic sphere often concluded with harsh resolutions, like advocating the use of violence against 'weak' and 'wicked' women" (2003: 166). Citing feminist scholarship that has discussed the gendered constitution of Hindu nationalism, Chaturvedi suggests that these discussions "were corollaries to the re-emergence of Maharashtrian brahmins in the 19th century, when forms of brahmanic Hinduism became important determinants for 'social' behavior, especially on the 'woman's question'" (166: 2003).

In Bengal, diagonally located across the sub-continent from Maharashtra, scholars have written prolifically on the manner in which woman becomes a key site for anti-colonial nationalists. In *Hindu Wife, Hindu Nation*, Tanika Sarkar writes about the fierce debates between Bengali nationalist-revivalists and the reformists about Hindu conjugality. Situated in a colonial context, the nationalist-revivalists sought to "critique colonialism", and the "loveless, purely deprivational unrequited nature of its political arrangement" through an alternate order – that of the Hindu household and the gendered relations between husband

and wife (2001: 39). Sarkar suggests that "Hindu nationalists needed to naturalise love as the basis for Hindu marriage, a higher form of love that excelled allegedly utilitarian, materialist and narrowly contractual Western arrangements" (2001: 40). So, Hindu marriage began to be idealized in 19th century Bengali novels and newspapers as "a complete spiritual union through perfect love" or a "lifetime of togetherness beginning with infancy" and "extending beyond the death of the husband" (2001: 40-41). In her life beyond the husband's death, the widow becomes "a reservoir of spirituality" for the nationalists (2001: 42). Or as Sarkar writes, "strict ritual observances root the widow's body in ancient India, thus miraculously enabling her to escape foreign domination. The cloth she wears is necessarily indigenous, the water she drinks is to be carried from the sacred river and not through foreign water pipes, and the salt that goes into her food is special rock salt untouched by machines. Ergo, the nation needs ascetic widowhood" (2001: 42). The ascetic qualities of Hinduism perceived in the practices of widowhood, illustrated a realm where the Bengali Hindu nationalists could assert their superiority. Self-immolation or the act of "sati" by widows on the funeral pyres of their husbands appeared an even better form of self-sacrifice (literally) for the Hindu nationalists. This act is also the name for the widow, the sati, who personifies self-sacrifice. Major literary figures such as Bankimchandra Chattopadhyay and Jyotirindranath Tagore saw the figure of the sati as a nationalist icon. Sarkar notes that "an immense body of patriotic tracts routinely invoked the act as an unfailing source of nationalist inspiration and pride" (2001: 43).[1]

According to Sarkar, these early Bengali Hindu nationalists saw the idealisation of the Hindu wife and widow as a way to oppose "reformist-cum-colonial-cum-missionary intervention into conjugality which had begun with renewed vigour from the 1870s – the Brahmo Marriage Act of 1873, proposals to introduce divorce in the 1880s, and the Age of Consent Act of 1891" (2001: 39). But the proposition that Hindu mar-

[1] The self-immolating widow, the sati, remains a problematic site in scholarship at the intersection of Indian feminist and subaltern studies (Spivak, 1988; Sarkar, 2002). In Gayatri Chakravarty Spivak's discussion, which traces the difficulty of conceptualising the subaltern in a colonial context, the sati figure shuttles between the colonial script of "white men saving brown women from brown men" and a nativist script which speaks of the desire of the sati to commit the act (1988). Sarkar critiques Spivak's position which, she suggests, does not take into account the narratives of Bengali women writing at the time. Sarkar suggests that the "longing for systematic learning", for example, "was not a desire implanted by male reformers, missionaries, and colonialism" (2001: 48). It is beyond the scope of this paper to discuss this literature in detail. However, it is necessary to assert, in this context, that sati was by no means an uncontested or an unregulated "tradition" even prior to British colonial governance (Daruwala, 1988).

riage was characterised by perfect love was untenable for a range of reasons that Sarkar outlines. Apart from the reformists who were engaged in pressuring colonial authorities to pass their legal proposals mentioned above, literary as well as legal challenges were posed by women themselves. Sarkar traces a "counter-tradition of folklore and verse that described marriage and domesticity as a source of profound unhappiness for the woman" (2001: 44). Since the 1860s, Bengali women writers critiqued the gendered arrangements of Hindu upper caste marriages. As one woman wrote, "not conjugal love nor children – could compensate for the deprivation of knowledge" and "yet foolish women serve them and care for them like servants" (2001: 47). Most women's writings, Sarkar suggests, identify the "pain of patrilocality and the longing for knowledge" (2001: 47). Legal challenges to an upper-caste attempt to idealise Hindu conjugality arose from cases such as that of Rukmabai in Maharashtra (Savarkar's linguistic region) who sought to have her marriage, contracted at infancy to a consumptive man, annulled (Sarkar, 2001: 48). Sarkar comments that "for some time, in Hindu nationalist circles, Rukmabai became a name more dreaded than Malabri's or those of the colonial legislators" (2001: 48). Malabri, as Sarkar notes, was a Maharashtrian social reformer who campaigned for a higher Age of Consent for marriage (2001: 50).

In the face of these challenges, Bengali upper-caste patriarchy began to rewrite conjugality in terms of "force and pain", often extolling the long-suffering virtues of the Hindu wife (Sarkar, 2001: 50). Yet this discourse appeared to have its limits even for the nationalists. Or as Sarkar argues, "the hegemonic desires of Hindu nationalism clashed too violently with the starkness of its discipline. Love had to re-enter the nationalist narrative" (2001: 51). This love took a different form. The ideal Hindu woman was abstracted into the "Motherland", a "new and supreme deity within the Hindu pantheon" (Sarkar, 2001: 51). Hence, in the Hindu nationalist context, debates about Hindu womanhood in terms of marriage and widowhood, and finally, motherhood, have played a key role in linking religion and nation – Mazzini's prescription for the glue that would hold the nation together.

Transmediterranean Dispersal II: Sonia Maino Gandhi and Hindu Nationalism

If the first transmediterranean dispersal has to do with Mazzini's influence on Hindu nationalism and its gendered constitutions, the second moment – in the scope of this discussion – has to do with the rise of Sonia Gandhi as a key political player in contemporary Indian politics. The dispersal in the second moment has to do with the ironic ways in which unity of religion and nation formulated by an Italian nationalist are deployed by Hindu nationalists against an Italian-born Indian national. Through this ironic dispersal, I discuss the event of Gandhi's election-victory as President of the Congress Party in 2004, and her subsequent refusal of the offer to become Prime Minister of India. I link this refusal to the Hindu nationalist opposition to her leadership which frames her as a foreigner and a Christian, unfit to take up the Prime Ministerial role. Furthermore, I discuss the manner in which Sonia Gandhi herself participates in a Hinduised gendered politics, which I argue, has always framed the manner in which even a "secular" party like Congress conducts its campaigns – a legacy of Sonia Gandhi's mother-in-law, Indira Gandhi (former Prime Minister of India).

In the 2004 elections, Sonia Gandhi was credited with leading the Congress party to electoral victory (Ninian 2004). Victory for United Provisional Alliance which the Congress Party led was a surprise as the Hindu Nationalist BJP and its allies in the form of the National Democratic Alliance were largely tipped to win (Butalia, 2004; Surendra, 2004; Ninian, 2004). In Surendra's (2004) analysis, the BJP's "India Shining" campaign, which used images of affluence to suggest its role in creating India's success through neoliberal economic reforms, had essentially backfired. "The 'India Shining' campaign", he states, "also told the truth about for whom India was shining for and whom the gains of the neoliberal economic reforms went – to the rich and mainly urban middle classes not the poor folk, lower castes and others" (2004: 85). Meanwhile the UPA had campaigned on the basis of *aam aadmi* or the common man. Yet, even as the UPA won in 2004, and the Congress Party voted for Sonia Gandhi to become Prime Minister of the country, Sonia Gandhi declined to become Prime Minister. The *Times of India* reported that Gandhi followed her "inner voice" and "humbly declined to take up the Prime Minister's post" (2004).

It would not be too difficult to assess whether it was Sonia Gandhi's inner voice or the very loud voices of Hindu nationalists which, in the end, persuaded Gandhi to give up the Prime Ministerial position. The BBC staidly reports that "her political opponents attempted to rake up her Italian descent as an election issue saying the choice for voters was between an Indian or foreign leader" (2009). But this was not a simple matter of the opposition raking up an issue. As Uwe Skoda states, "High drama followed the unexpected victory" (2004: 273). In the following account, Skoda details this drama:

> The prospect of the Italian born leader Sonia Gandhi… becoming Prime Minister (PM) of India led to a furious agitation by prominent politicians of the losing Bharatiya Janata Party (BJP). Madhya Pradesh Chief Minister Uma Bharati staged a mock resignation, party ideologue Govindacharya staged a nation-wide campaign against the *videshi bahu* (foreign daughter-

> in-law), and Parliamentary Affairs Minister Sushma Swaraj… declared she would tonsure her head and stay in mourning in a case a 'foreigner" became PM. Of course, Sonia, after a heated national and internal party debate, finally decided against seeking the office of PM, nominating INC stalwart and former Finance Minister Manmohan Singh instead. (2004: 273)

The emotions expressed in these threats by the opposition are enough to rival any Bollywood scenario. I can even hear the cliché from the tongue of a Bollywood patriarch or matriarch when confronted with an unthinkable transgression to the familial social order: *Nahi, Yeh Kabhi Nahi Ho Sakta* (No, This Can Never Happen).

Thinking through an unthinkable event is often necessary from a deconstructivist perspective. In an opinion piece in the *New Internationalist*, Urvashi Butalia does precisely this. Marking the links between skin-colour, nationality, citizenship, allegiances, choice of food and clothes made in the case of opposition to Sonia Gandhi's Prime Ministership, Butalia analyses the flip side of such links. "Indians are in every single country in the world today", she states, "They're making their fortunes, fighting for their rights, and they want to be seen as citizens of the country they have made their home… How then can we castigate anyone who makes our country their home" (2004: 5). Commenting further on the gendered nature of Sonia Gandhi's relationship to the Gandhi family, Butalia looks at the irony of Hindu nationalist designation of Sonia Gandhi as a foreigner. "By marrying Rajiv Gandhi", she states, "she gave up claim to any previously held identity (indeed the BJP would be the first to say this). How does it suddenly become possible for the woman to remain an autonomous subject when all her life she has been (supposedly) secondary to the man?" (2004: 5). Butalia's analysis hits at the core of the manner in which Hindu nationalism functions. The links between race, gender, colour, nationality, citizenship and allegiances of religion or food or clothing often unravel when ironies such as the ones above are noted. What of the status of the precious Non-Resident Indians (especially the successful ones) who were being claimed by the BJP during its rule through the declaration of a special day – Pravasi Bharatiya Divas – for Overseas Indians.[2] Indians abroad were being validated and welcomed by the Indian government not least for their financial ability to invest in India's economy. And what of the gendered politics of the Hindu wife, the loyal and chaste widow, the dutiful mother? Didn't Sonia Gandhi embody all these qualities? Not only did she embody these qualities, she actively constructed herself as an exemplary Hindu wife (as I discuss below) – but that hasn't been enough. For that moment, however, Sonia Gandhi took

2 See http://www.pbdindia.org/.

the wind out the sails of the opposition by declining to take up the role of Prime Minister.

In defusing the opposition, Butalia suggests that Sonia Gandhi "put to rest, at least for the time being, a furious debate on nationality and citizenship" (2004: 5). But even though "hundreds of millions of voters in India" voted for Gandhi, as Butalia states, many Hindutva groups, continue to highlight Sonia Gandhi's Italian origins and her Catholic faith as marks of her non-Indianness (2004: 5). For example, in 2007, US based Hindutva groups staged a protest against Sonia Gandhi, who was invited by the United Nations to speak on Mahatma Gandhi's birthday, suggesting she was an inappropriate representative. *The Hindustan Times* stated that the "protest has been organised by the New York based Forum for Saving Gandhi Heritage which seems to have been recently floated for this protest" (2007). Furthermore, Hindutva groups like the Vishwa Hindu Parishad, continually frame her Christian identity as evidence of anti-nationalism. In 2008, President of the Vishwa Hindu Parishad called for an investigation into an alleged link between Sonia Gandhi and Opus Dei. As the Indo-Asian news service reports: "Describing Sonia Gandhi as the 'daughter of Vatican' and Opus Dei, the 'global Catholic outfit' that has an anti-India agenda, VHP president Ashok Singhal said: 'Investigations must be made into the relationship between Sonia Gandhi and the Opus Dei and in this light, investigations must also be made to find the mastermind behind the unusual death of Sanjay Gandhi, Indira Gandhi and Rajiv Gandhi'" (2007). The VHP's anti-conversion agenda is apparent in this statement, as it represents a Catholic organisation as an anti-Indian one simply because of its potential for conversion. The anti-conversion position, in this context, is used to make the allegation that somehow Sonia Gandhi is responsible for all the tragic deaths of the Gandhi family. While such an allegation seems absurd, the VHP's active anti-conversion campaigns, their long-term activism to foment violence against Muslims and Christians is not. Angana Chatterjee, for instance, has described the activism of the VHP in Orissa over the last decade which has seen recent violence against tribal Christian communities in the area (2004: 326-328). In Hindu nationalist arguments, therefore, Sonia Gandhi is unable to represent the Indian nation because she is of "foreign origin" and she is "Catholic". In this argument, Indian Christians are necessarily positioned either as dupes who have lost their Indianness or as traitors to the nation.

Beyond the mobilization of Hindu nationalism against Sonia Gandhi, however, another question needs to be asked. What of Sonia Gandhi's own participation in a gendered Hinduised politics? As Skoda's research on the 1999 elections reveals, Gandhi's entry into Indian politics was made on the basis of positioning herself as a good Hindu wife, widow, and mother – a daughter-in-law of the nation. Skoda notes that Sonia Gandhi "acted in accordance with the traditional behavior expected of a widow" and "often covered herself in a *dupatta* (loose veil) and had her election posters painted accordingly" (2004: 280). In her speeches, she referenced "Mother Indira" and "husband Rajiv", thereby attempting to convince voters that "as a good *bahu*, she was still completely loyal to her husband and his legacy" (Skoda, 2004: 281). Gandhi had "decided to join active politics only after 'people had begged me with so much love, I had to do it for them and the family'" (Skoda, 2004: 281). While it may be true that Sonia Gandhi did join politics to raise the flagging fortunes of the Congress Party, in such a representation, she positions herself as the self-sacrificing widow who gives all for the family, the party and the nation.

In this political strategy, Sonia Gandhi follows the footsteps of another master strategist in the use of gendered discourses, her famous mother-in-law, Indira Gandhi. Indira Gandhi's representation of herself as an embodiment of *Bharat Mata* or India's Mother was based on her widowed mother status – a status rich with virtuous and nationalist resonances within anti-colonial nationalist literature as well as popular cultural forms such as Bollywood films. This embodiment was her campaign persona. She often addressed campaign rallies dressed in sober widow white and bordered *khadi* (handwoven) sarees, with her head covered to signify her chastity. Skoda goes on to argue that Sonia Gandhi's positioning of herself as the ideal Hindu wife, widow, and mother was based on Indira Gandhi's strategies. Yet, she constructs an anthropological representation of gendered kinship relationships in rural North India to explain the power of these strategies. Or as she states, "fundamentally important in understanding the semantics of both *beti* and *bahu* categories in north India is the widely prevalent rule of local, especially village, exogamy" (2004: 275). Such a representation of *beti* (daughter) and *bahu* (daughter-in-law) flattens these relationships – without an eye to how many of these relationships may be lived in practice. For example, the economic status of a given relative may inflect those relationships quite differently. Secondly, there is a lack of reference to the manner in which a particular North Indian rural practice is not necessarily practised in all Indian communities. Thirdly, to suggest that gendered kinship relations continue across time without any changes disavows the political contexts within which "traditions" might be invented especially with regard to the context of Hindu womanhood.

As I discussed earlier, Hindu nationalist debates positioned an ideal Hindu woman as key to a Hinduised nation. Indian womanhood – often represented as Hindu womanhood – has had a long history of representation in Hindi commercial cinema or Bollywood. In her seminal book, *National Identity in Indian Popular Cinema 1947-1987*, Sumita Chakravarty discusses the manner in which the 1957 Bollywood classic, *Mother India* represents the Indian mother as possessing "mythic strength and endurance" and shows her "ability to provide stability and continuity in Indian social life" (1993: 150). *Mother India's* director, Mehboob states, "I have always had the most profound respect for Indian womanhood" (quoted in Chakravarty, 1993: 150). This respect appears to have been "centred around the fact that the true Indian woman enters her husband's home when she marries and leaves it only when she dies, that she will never sell her chastity for any price on earth" (quoted in Chakravarty, 1993: 15). This attitude to Indian womanhood, as Chakravarty notes, is expressed through the film's "chronicle of one woman's struggle against the oppressions of both man and nature", but it also becomes an "encapsulation of India's long history of domination by foreign powers and its struggle to maintain the integrity of its soil" (1993: 151). This representation of Indian womanhood, its resonance of austerity and chastity, for example, might have made the early Hindu nationalists proud. Quite accurately, Chakravarty links this representation of Indian womanhood to the manner in which Indira Gandhi, who was elected India's Prime Minister in 1967 (9 years after the release of *Mother India*), represented herself. Indira Gandhi "was cheered as *Bharat Mata*, an association she herself encouraged" after India's war against Pakistan in 1971 (1993: 152).

Growing up in India, among a Congress-I voting family, my childhood brain associated Indira Gandhi with the actual map of India, as if the *khadi* or handloom sarees that she wore to woo the electorate, elegantly shaped across her hips, shaped India. Her strategy to represent herself as the widowed mother of the nation succeeded in this respect. In the 1970s, I remember posters which depicted her along with three young children representing the army, the navy and the airforce. I thought of those branches of the nation's defence as her children. Her assassination came as something of a shock to me when I was 16 years of age. I remember thinking that there was no one who could lead the nation – that somehow chaos was to follow. Violence did – in the form of vengeance against Sikh communities especially in Delhi where I lived. But Indira Gandhi's death was not the end of the Gandhi family's political power. Her son Rajiv Gandhi held Prime Ministerial office from 1984-1989. He was leader of the Congress Party until 1991, when he was assassinated by the *Liberation Tigers for Tamil Ealam* while campaigning for elections in the state of Tamil Nadu. This event, albeit tragic, has been used strategically to reinvigorate the fortunes of the Congress Party and the Gandhi family. In this sense, Sonia Gandhi can be read as the clever, strategic daughter-in-law, drawing on her mother-in-law's kitchen politics strategies, and participating in a Hinduised politics (at least in its gendered aspect) which a secular Congress party supposedly differentiates itself from.

In reading this Hinduised politics that the Congress Party engages in, it is necessary to note that Skoda's reading of Sonia Gandhi also Hinduises India through her construction of the nexus between Hindu kinship relations and Indian politics. As she states, "to both act within the framework of kinship roles and emphasize dynastic relations in an attempt to win political office point to a clear nexus between politics and kinship in India" (2004: 283). The kinship relations which she references are linked to North Indian rural Hindu relationships. In arguing for such a reading, she further entrenches a dominant idea of the unity of religion and nation-state that a Hinduised politics assumes. Furthermore, her reading of Indian politics appears to be Orientalist one. It would be good to remind ourselves that political families as well as gendered, marital relationships do play a role in politics in the West. The Kennedy family, for example, still holds electoral clout in the US through various associated members and descendants. Furthermore, representations of normative families hold a great deal of power in election campaigns in the US. Instrumental uses of family members in campaigns for political leadership play a major role. Family scandals become national and transnational scandals. US President Bill Clinton's affair with Monica Lewinsky became an ongoing national and transnational media affair. Hillary Clinton's decision to "stand by her man" fitted in with a conservative gendered attitude to the sanctity of the family. One could speculate, for instance, that her decision was made in the interest of her own political aspirations. But the fact remains that the public representation of the marital relationship was scripted by conservative, gendered familial norms of the strong and forgiving wife and the repentant husband which seemed to appeal to the American public. The representation of a stable African American First Family whose only initial problem was finding an appropriate puppy – a gift for the girls who endured their father's demanding campaign – illustrates US President Barack Obama's media-savvy cool. In a nation where African American families are often represented as dysfunctional or non-normative (the single mother on welfare stereotype), the intelligent, good-looking, fashionable Cosby-like First Family has gained celebrity status in the global mediascape. Drawing on heteronormative, familial norms to advance political careers, therefore, is not unique to India's electoralscape. By foregrounding an Orientalist and naturalised Hindu representation of kinship relations in Indian politics, Skoda misses the point that the campaign framing of Sonia Gandhi as good or bad Hindu daughter-in-law, in fact, worked within the parameters of Hindu nationalism's constructed gendered links between religion and nation.

The distinctiveness of Sonia Gandhi's story lies, I would argue, not in a subcontinental performance of kinship relation through politics, but in her Italian ancestry and its significance in the context of a Hinduised politics of Indian elections. At the intersection of Hindu nationalist representations of Sonia Gandhi and her own participation in a Hinduised gendered politics lies the fact Sonia Gandhi is an Italian-Indian Catholic political leader who married into a cosmopolitan secular family, whose anti-colonial commitment and credentials still carry weight, rightly or wrongly, for the Indian citizenry. As Butalia reminds us, "hundreds of millions of voters" voted for Sonia Gandhi (2004: 5). As an Italian-Indian, Sonia Gandhi embodies a transmediterranean dispersal – through marriage and migration. Despite the influence of an Italian religion-based nationalism which has been localised in the Indian context, she is a political leader who unsettles the unity of religion, race, skin-colour, and nationalism because of her Italian connection. Through this transnational connection, her embodiment as an Italian Indian poses a challenge to that earlier Mazzinian formulation of nationalism. Sonia Gandhi's assertion of Indianness – "I never felt they look at me as a foreigner. I am not. I am Indian" (BBC 2009) – needs to be placed in the context of an Indianness which is not reduced to either Hinduness or a stereotypical idea of Indianness (one that argues for distinctions between cultural or racial indigeneity and non-indigeneity). In a country whose metropolitan cities boast of eateries which advertise pizzas and pastas, Indianness cannot be anchored to a homogenous idea of culture or nation. This is the lesson that a mapping of transmediterranean dispersals has to offer us. It makes visible a scaffolding of nationalism wherein indigeneity and non-indigeneity are politicised descriptions rather than self-evident markers.

Epilogue

Joseph PUGLIESE

Associate Professor at Macquarie University, Sydney

In the wake of the transmediterranean crossings staged by the various contributors to this book, a number of questions and issues have been brought into focus. These questions and issues largely pivot on a geopolitical divide that has lost none of its force or power: Global North/South. As a globalised division, this geopolitical marker is instrumental in the sorting of the subjects who inhabit the Mediterranean region, in all of its dispersed manifestations. This geopolitical division is a constant burden throughout the majority of the essays in this collection. Importantly, a number of the contributors proceed to situate this geopolitical division in an historical context, precisely as they mark its contemporary effects and embodiments. Enforced deportations, migrants, refugees, asylum seekers, "unauthorised" arrivals, *extracommunitari*, aliens and so on – this catalogue bears testimony to subjects compelled to embody the figure of the subaltern who is subject to the exercise of legislated violence by the nation-state.

Inscribing so many of the histories that are traced by the essays in this volume is the haunting and still very active spectre of colonialism. The Mediterranean emerges, in the pages of this volume, as a geopolitical space harrowed by stratified histories of colonialism. The contemporary conflict in Darfur is shown to be underpinned by a complex mesh of colonial and racist palimpsests that configured bitter divisions between ethnicities that continue to ramify. The legacy of Europe's crimes against its Jewish peoples continues to shape the lives and deaths of contemporary Palestinians. Contemporary carceral archipelagos in both the Mediterranean and off the coast of the Australian coast are shown to be enabled by prior histories of colonial rule. Questions of sovereignty in the context of unresolved colonial histories continue to be raised in the face of contemporary nations such as Italy and Australia that declare themselves to be "postcolonial" states, when, in fact, there has been no successful overcoming of colonial history or relations.

In the face of the problematics and irresolutions that are systematically documented across the essays in this volume, something else continues to persist: the affirmation of the figure of *trans* that insists on interrogating and disrupting the conceptual boundaries of received histories and the delimiting borders of statist thinking. The animating figure of *trans* is eloquently embodied by Said's conceptualisation of a politics of "affiliation" (discussed by Shalbak, this volume), as a politics that refuses the homogenised categories so useful to the violent, biopolitical operations of the nation-state. The transitive movement of affiliation affirms, instead, a crossing of categories (of nation, ethnicity, religion and so on) that enunciates difference and alterity in the face of the imperialism of the same.

The essays in this volume bespeak differences, in the plural. A polyglossia of Englishes articulates the embodied differences of subjects inscribed by complex diasporic histories. Through this linguistic medium, a singular language is reaccentuated and resignified through subjects that have known, as lived experience, the diasporic processes of dispersal and displacement. Speaking back to the Mediterranean, with this multi-inflected language and from positions of displacement, the contributors to this volume enunciate the possibility of a Mediterranean that is not identical to itself (as if it ever was); they delineate a *transmediterranean* that has effectively been transposed beyond its geographic borders to other spaces altogether. The transitive movements that score this volume establish lines of connection between Africa, Europe, Asia and Oceania, establishing, in the process, a *circoscrizione mostro* or a "monster constituency". I co-opt this teratological term from Vanni's essay (this volume) in order to affirm an exorbitant collectivity that defies the disciplinary relations and categorical limits of the nation-state in order to gesture to alternative ways of thinking and acting in the world. This is not, I emphasise, a globalised constituency. On the contrary, it is a constituency that locates itself in the interstices of the global, articulating fragile connections and tactical relations that critically interrogate the homogenising and normative violences of the global. This monster constituency, that traduces the Mediterranean of the imperial *mare nostrum* (our sea) into an emanation that cannot claim the originary ground of the proper (*locus proprius*), discloses, in the process, what would otherwise be unspeakable: other histories, stratified colonialisms, subaltern subjects and embodied differences that, in the face of multiple silences and dispersals, are magnetised into being.

On the edges of the Mediterranean, a series of voices is mobilised. Heterological in their embodiments, teratological in their constituency, these voices transpose what was proper, normative and seemingly identical to itself to a dispersed domain of overlapping spaces, buried histories and dislocated subjects. A heterogeneous cluster of names – Palombo, Said, Doukas, ʻAbd al-Rahman, Cilento, Giannacopoulos, Gandhi, Dalkin, Osuri, Alexiou, Shalbak, and so on – is transposed to a scattered geography of spaces – Suez, Darfur, Lampedusa, Aivali, New Delhi, Five Dock, Himadio, Palestine, and so on. In the wake of these multiple transpositions, something significant remains: a tenacious affirmation of transmediterranean affiliations on altered ground.

Contributors

Petro Alexiou is affiliated with Macquarie University. His PhD was based on a critical biography of the Greek diasporic writer Alekos Doukas. His research focus is on Greek diaspora and the cultures of the Eastern Mediterranean. His work has been published across a wide range of international journals.

Noah Bassil has a PhD in Politics and International relations on the topic of the conflict in Darfur, western Sudan. The project grew out of an interest in the question of identity politics and the formative questions of the political economy of Sudan, such as inequality, development and underdevelopment, and conceptualizations of the post-colonial state in the Middle East and Africa. He is Deputy Director of the Macquarie University Centre for Middle East and North African Studies and has published on the topics of Darfur, colonial racial discourse and post-colonial racism, and Arab and African identities.

Seren Dalkiran is a student in Sociology and Anthropology with a specific interest in gender, ethnicity and social justice at the International Honours College, University College Utrecht. She is also a documentary maker and the co-founder of the national foundation "Lost Faces" in the Netherlands that strives for human rights, particularly gender equality and empowerment of youth.

Maria Giannacopoulos is affiliated with Macquarie University. She has a PhD in Critical and Cultural Studies. Her published research is concerned with law's violence specifically in relation to Australia's racial history. Her work has been published in both books and international journals.

Gaia Giuliani teaches in the Department of Politics, Institutions and History, University of Bologna. She has published in Italian the translated work of Judith Butler, and has published her work in both books and international journals. She is on the editorial board of *Studi Culturali*, and she is a recipient of an Australian Endeavour Postdoctoral Fellowship.

Goldie Osuri teaches in the Department of Media, Music and Cultural Studies, Macquarie University. Her research is focused on diasporic Indian cultures, whiteness studies, and religious conversion in colonial contexts. Her research has been published in a number of international journals, including *Social Semiotics, Cultural Studies*

Review and *Media, Culture and Society*. She is co-editor of *Borderlands*.

Lara Palombo (Hons., M.A.) is affiliated with Macquarie University, and is a PhD scholar researching the internment of women of Italian origins during the Second World War. She has published across a range of international journals.

Joseph Pugliese is an Associate Professor in Critical and Cultural Studies. He teaches in the Department of Media, Music and Cultural Studies, Macquarie University. He researches on race and ethnicity, whiteness, migration and diaspora, refugees and asylum seekers, bodies and technologies, and cultural studies of law. His research has been published across a wide range of books and international journals. His most recent publication is a monograph, *Biometrics: Bodies, Technologies, Biopolitics* (New York, Routledge, forthcoming 2010), that examines the biopolitics of biometric technologies in both historical and contemporary contexts.

Ihab Shalbak is a PhD candidate at the University of New South Wales, Sydney. He is researching the role of think tanks in the context of transnational politics. He is a research specialist on contemporary Palestine and Edward Said.

Ilaria Vanni teaches in the Department of International and Cultural Studies, University of Technology, Sydney. She has published in international journals on Italian diasporic culture, contemporary art, and Australian Aboriginal art. She is also a curator who has curated exhibitions at the Museum of Sydney, and at the National Maritime Museum, Sydney.

References

'ATHENIAN' (1972), *Inside the Colonels' Greece*, trans. R. Clogg, London, Chatto and Windus.

ABC NEWS ONLINE (2006), "Greeks 'a wonderful example of integration,'" http://www.abc.net.au/news/stories/2006/09/15/1742129.htm.

ABDIN, H. (1985), *Early Sudanese Nationalism, 1919-1925*, Khartoum, Khartoum University Press, 1985.

ABU-LUGHOD, L. (2002), "Do Muslim Women Really Need Saving? Anthropological Reflections on Cultural relativism and Its Others", *American Anthropologist*, Vol. 104, No. 3.

ADAS, M. (1989), *Machines as the Measure of Men: Science, Technology and Ideologies of Western Dominance*, Ithaca, Cornell University Press.

AGAMBEN, G. (1998), *Homo Sacer: Sovereign Power and Bare Life*, trans. D. Heller-Roazen, Stanford, Stanford University Press.

AGAMBEN, G. (2005), *State of Exception*, Chicago, University of Chicago Press.

AIKEN, H. D. (1956), *The Age of Ideology*, New York, A Mentor Book.

ALEXAKIS, E. and JANISZEWSKI, L. (1998), *In Their Own Image: Greek Australians*, Sydney, Hale and Iremonger.

ALEXIUO, P. (2005), "Diaspora and Colonialism in Australia in the 1920s: the case of Alekos Doukas's migrant 'voyage south'", *Modern Greek Studies Australia & New Zealand*, Vol. 13.

ALEXIUO, P. (2008), "A Body Broken: A critical biography of Alekos Doukas (1900-1962)", PhD Thesis, Department of Critical and Cultural Studies, Macquarie University, NSW.

ALOMES, S. and JONES, C. (1991), *Australian nationalism. A Documentary History*, North Ryde, HarperCollins.

AMIN, S. (1972), "Underdevelopment and Dependence in Black Africa: Historical Origin", *Journal of Peace Research*, Vol. 9.

AMNESTY INTERNATIONAL REPORT (2009), www.report2009.amnesty.org/en/regions/europe-central-asia/italy.

ANDERSON, B. ([1983] 1991), *Imagined Communities*, London and New York, Verso.

ANDERSON, K. (2007), *Race and the Crisis of Humanism*, London and New York, Routledge.

ANDERSON, W. (2002), *The Cultivation of Whiteness. Science, Health, and Racial Destiny in Australia*, Durham, Duke University Press.

ANESTIDIS, S. ([1889] 2002), "Introduction", in I. I. KALFOGLOUS (ed.), *Istoriki Geografia tis Mikrasiatikis Xersonisou* (Historical Geography of the Asia Minor Peninsula), Athens, Kentro Mikrasiatikon Spoudon.

ANG, I. (2001), *On Not Speaking Chinese*, London, Routledge.

APOSTOLAKIS, I. D. (1914), *Kidoniakai Meletai kai Parallilismoi* (Kydonies Studies and Parallels), Aivali, Typografeion 'O Ilios'.

APPADURAI, A. (1996), *Modernity at Large, Cultural Dimensions of Globalization*, Minneapolis and London, University of Minnesota Press.

APTER, A. (1999), "Africa, Empire, and Anthropology, A Philological Exploration of Anthropology's Heart of Darkness", *Annual Review of Anthropology*, Vol. 28.

ARENDT, H. (1951), *The Origins of Totalitarianism*, New York, Harcourt, 1968.

ARGIROPOULOU, R. (1995), *H Filosofiki Skepsi stin Ellada apo to 1828 os to 1922: Evropaikes epidraseis kai prospatheies gia mia ethniki filosofia 1828-1875* (Philosophical Thought in Greece from 1828 to 1922: European Influences and Attempts for a National Philosophy 1828-1875), Vol. 1, Athens, Gnosi.

ARMSTRONG, J. (1976), "Mobilised and Proletarian Diasporas", *American Political Science Review*, Vol. 70.

ASAD, T. (1970), *The Kababish Arabs: Power, Authority and Consent in a Nomadic Tribe*, London, C. Hurst and Company.

AUSTEN, R. A. and WOODRUFF, D. S. (1969), "Images of Africa and British Slave-Trade Abolition: The Transition to an Imperialist Ideology, 1787-1807", *African Historical Studies*, Vol. 2, No. 1.

AVDELA, E. (1998), "Class, Ethnicity, and Gender in Post-Ottoman Thessaloniki: The Great Tobacco Strike of 1914", in B. MELMAN (ed.), *Borderlines: Genders, identities in war and peace 1870-1930*, New York, London, Routledge.

BAGARANI, M., BONETTI, A. and ZAMPINO, S. (2007), "Decentralization of Territorial Policy in Italy – the Coherence with the Model of Multi-level Governance and the Effects on Responsibilities of Public Spending", EconPapers, http://econpapers.repec.org/paper/molecsdps/esdp07038.htm.

BASHFORD, A. (2004), *Imperial Hygiene. A Critical History of Colonialism, Nationalism and Public Health*, London, Palgrave.

BATALOVA, J. (2008), "Immigrants in the US Armed Forces", in *Migration Information Source*, www.migrationinformation.org/feature/print.cfm?ID=683.

BATTISTON, S. and MASCITELLI, B. (2008), *The Italian expatriate vote in Australia. Democratic right, democratic wrong or political opportunism?*, Ballan, Connor Court Publishing.

BAUBOCK, R. (2005), "Expansive Citizenship-Voting Beyond Territory and Membership", *Political Science & Politics*, Vol. 38, No. 4.

BAUBOCK, R. (2007), "Stakeholder Citizenship and Transnational Political Participation: A Normative Evaluation of External Voting", *Fordham Law Review*, Vol. 75, No. 5.

BAUBOCK, R. (2003), "Towards a Political Theory of Migrant Transnationalism", *The International Migration Review*, Vol. 37, No. 3.

BBC NEWS (2009), "Profile: Sonia Gandhi", 30 March, http://news.bbc.co.uk/1/hi/world/south_asia/3546851.stm.

BBC NEWS (2008), "Italy fingerprint plan criticized", BBC News, 26 June 2008, http://news.bbc.co.uk/2/hi/europe/7476413.stm.

BEBBINGTON, A. (2000), "Reencountering Development: Livelihood Transitions and Place Transformations in the Andes", *Annals of the Association of American Geographers*, Vol. 90.

BENJAMIN, W. ([1973] 1992), *Illuminations*, London, Fontana Press.

BESHIR, M. O. (1974), *Revolution and Nationalism in the Sudan*, London, Rex Collins Ltd.

BHABHA, H. (2004), 'Untimely Ends', *Artforum International*, Vol. 42, No. 6.

BIRCH, T. (2000), "The Last Refuge of the 'Un-Australian,'" Transforming Cultures Winter School, www.transforming.cultures.uts.edu.au/pdfs/last_refuge_birch.pdf.

BODDY, J. (2008), "Anthropology and the Civilizing Mission in Colonial Sudan", Firth Lecture, London School of Economics, http://www.lse.ac.uk/collections/anthropology/pdf/asa_conf_08/Boddy_Firth_Lecture_2008.pdf.

BOURDIEU, P. (1971), "Systems of Education and Systems of Thought", in YOUNG, M. F. Young (ed.), *Knowledge and Control: New Directions for the Sociology of Education*, London.

BRANDABUR, C. A. and ATHAMNEH, N. A.H. (2000), "Problems of Genre in The Seven Pillars of Wisdom: A Triumph", *Comparative Literature*, Vol. 52, No. 4.

BULL, M. and Newall, J. (2005), *Italian Politics: Adjustment Under Duress*, Cambridge, Polity Press.

CARTER, D. (1985), "Re-Viewing Communism: *Communist Review* (Sydney), 1934-1966: A Checklist of Literary Material", *Australian Literary Studies*, Vol. 12, No. 1.

CARTER, D. (1989), "Biography, Politics, a Novel: Reading Judah Waten", *Southern Review*, Vol. 22.

CARTER, D. (2002), "Communism and Carnival: Ralp de Boissiere's Crown Jewel and its Australian context", *Australian Cultural History*, Vol. 21.

CASSANELLI, L. V. (1988), "The Ending of Slavery in Italian Somalia", in MIERS, S. and ROBERTS, R. (eds.) *The Ending of Slavery in Africa*, Madison, The University of Wisconsin.

CHATERJEE, A. (2004), "The Biopolitics of Hindu Nationalism: Mournings", *Cultural Dynamics*, Vol. 16.

CHATURVEDI, V. (2003), "Vinayak & Me: Hindutva and the Politics of Naming", *Social History*, Vol. 28, No. 2.

CHATZEIOSEPH, C. (2002), "To prosfigiko sok, oi statheres kai oi metavoles tis ellinikis oikonomias" (The Refugee Shock, the constants and variables of the Greek Economy), in CHATZEIOSEPH, C. (ed.), *The History of Greece in the 20th Century: 1922-1940 The Interwar*, Vol. 2 (Part 2), Athens, Vivliorama.

CHATZEIOSEPH, C. (2003), "Koinovoulio kai Diktatoria" (Parliament and Dictatorship), in CHATZEIOSEPH, C. (ed.), *The History of Greece in the 20th Century: 1922-1940 The Interwar*, Vol. 2 (Part 2), Athens, Vivliorama.

CHATZEIOSEPH, C. (ed.), *The History of Greece in the 20th century: 1922-1940 The Interwar*, Vol. 2 (Part 1), Athens, Vivliorama.

CHAUDHARY, A.S. (2005), "The Simulacra of Morality: Islamic Veiling, Religious Politics and the Limits of Liberalism", *Dialectical Anthropology*, Volume 29, No. 3-4.

CHESTERMAN, J. and GALLIGAN, B. (1997), *Citizens without Rights*, Cambridge, Cambridge University Press.

CHOUIRMOUZIOS, A. (1940), "Logotechniki Alitografia" (Vagabond Literature), *Nea Estia*, Vol. 27. No. 313.

CHOW, R. (1990), *Writing Diaspora*, Bloomington, Indiana University Press.

CHOW, R. (2002), *The Protestant Ethnic and the Spirit of Capitalism*, New York, Columbia University Press.

CHOWDHRY, G. (1995), "Engendering Development? Women in Development (WID) in International Development Regimes", in MARCHAND, M. and PARPART, J. (ed.), *Feminism/Postmodernism*, New York, Routledge.

CHRISTMAS ISLAND TOURIST ASSOCIATION (2009), Christmas Island Tourist Association Official Website, http://www.christmas.net.au/introduction.php.

CILENTO, R.W. (1925), *The White Man in the Tropics, with Especial Reference to Australia and Its Dependencies*, Melbourne, Government Printer.

CILENTO, R.W. (1944), *Blueprint For the Health of a Nation*, Sydney, Scotow Press.

CILENTO, R.W. (1959), *Triumph in the Tropics. An Historical Sketch of Queensland*, Brisbane, Smith & Patterson.

CLARK, M. (1984), *Modern Italy 1871-1982*, London, Longman.

CLIFFORD, J. (1997), *Routes: travel and translation in the late 20th century*, Cambridge, Mass., Harvard University Press.

CLOGG, R. (1972), "Two Accounts of the Academy of Ayvalik (Kydonies) in 1818-1819", *Revue des Études Sud-Est Européennes*, Vol. X, No. 4.

CLOGG, R. (1992), *A Concise History of Greece*, Cambridge, Cambridge University Press.

COLUCCI, M. (2002), "Voto degli italiani all'estero", in P. BEVILACQUA, A. DE CLEMENTI, E. FRANZINA (eds.), *Storia dell'emigrazione italiana*, Arrivi, Roma, Donzelli Editore.

COMMONWEALTH OF AUSTRALIA. (2007), *Becoming an Australian Citizen*, Canberra.

CUNSOLO, R. S. (1990), *Italian Nationalism. From its Origins to Second World War*, Malabar, Florida, Robert E. Krieger Publishing Company.

CURTHOYS, A. (1999), "Expulsion, Exodus and Exile in White Australian Historical Mythology", *Journal of Australian Studies*, Vol. 61.

DALY, M.W. (1986), *Empire on the Nile: The Anglo-Egyptian Sudan, 1898-1934*, Cambridge, Cambridge University Press.

DALY, M.W. (2007), *Darfur's Sorrow: A History of Destruction and Genocide*, Cambridge, Cambridge University Press.

DARRAJ, F. (2002), *The Memory of the Conquered: Defeat and Zionism in the Palestinian Cultural Discourse*, Casablanca, The Arabic Cultural Centre.

DARUWALA, M. (2009), "Central Sati Act – An Analysis", *South Asian Citizens Web*, 15 May 2009.

DARWISH, M. (2004), *Edward Said: A Contrapuntal Reading*, Cairo, Al-Ahram Press.

DAVIS, D. (2007), "Yo Soy el Army", in *Metroactive*, http://www.metroactive.com/metro/09.19.07/news-0738.html.

DE CERTEAU, M. (1984), *The Practice of Everyday Life*, Berkeley, University of California Press.

DE WAAL, A. (2005), "Who are the Darfurians? Arab and African identities, violence and external engagement", *African Affairs*, Vol. 104, No. 415.

DEER, P., PRAKASH, G. and SHOAT, E. (2006), "Edward Said: A Memorial Issue", *Social Text*, Vol. 24, No. 2.

DENG, F. M. (1995), *War of Visions: conflict of identities in the Sudan*, Washington, D.C., Brookings Institution.

DEPARTMENT OF IMMIGRATION AND MULTICULTURAL AFFAIRS [DIMA]. (2006), "Australia to Introduce Citizenship Test", Media Release 11 December, http://www.minister.immi.gov.au/parlsec/media/ mediareleases/ medrel06/111206.htm.

DERRIDA, J. (1976), *Of Grammatology*, Baltimore and London, The Johns Hopkins University Press.

DERRIDA, J. (1992), "Force of Law: The 'Mystical Foundation of Authority'", in CORNELL, D., ROSENFELD, M. and CARLSON, D. G. (eds.), *Deconstruction and the Possibility of Justice*, New York, Routledge.

DIRK, N. (2004), 'Edward Said and Anthropology', *Journal of Palestine Studies*, Vol. 33, No. 3.

DOUGLASS, W. (1999), *Italians in North Queensland*, Brisbane, University of Queensland Press.

DOUKAS, A. (1953), *Stin Pali Sta Neiata* (To Struggle, To Youth), Melbourne, Coronation Press.

DOUKAS, A. (1963), *Kato Apo Ksenous Ouranous* (Under Foreign Skies), Melbourne, Australian Greek Publications.

DOUKAS, S. (1976), *Enthimimata apo Deka Filous mou* (Remembrances of Ten Friends), Athens, Kedros.

DOVEY, K. (1993), "Rottnest Island: Rust and Irony", *Meanjin*, Vol. 21.

DRAKOS, E. (1895), *Ta Mikrasiana: I Geniki Pragmateia Peri Ekatonison Tanin Moschonision* (Asia Minor Matters: A General Treatise Regarding Hecatonesoi. Moschonisia today), Issue One, Athens,

DUNNE, T., KURKI, M. and SMITH. S. (2007), *Theories of International Relations. Discipline and Diversity*, New York, Oxford University Press.

DYER-WHITEFORD, N. (1999), *Cyber-Marx: Cycle and circuits of struggle in high technology capitalism*, Urbana and Chicago, University of Illinois Press.

EDKINS, J. and PIN-FAT, V. (2004), "Life, Power, Resistance", in EDKINS, J., PIN-FAT, V. and SHAPIRO, M. J. (eds.), *Sovereign Lives: Power in Global Politics*, London, Routledge.

EDKINS, J. and PIN-FAT, V. (2005), "Through the Wire: Relations of Power and Relations of Violence", *Millennium: Journal of International Studies*, Vol. 34, No. 1.

EDKINS, J., PERSRAM, N. and PIN-FAT, V. (eds.). (1999), *Sovereignty and Subjectivity*, Boulder, L. Rienner.

EDWARDS, P. (1997), *A Nation at War: Australian Politics, Society and Diplomacy during the Vietnam War 1965-1975*, Sydney, Allen and Unwin.

EL-AMIN, M. N. (1992), "Was there an Alliance between the Watanist (Nationalist) Party, International Communism, and the White Flag League in the Sudan", *British Journal of Middle Eastern Studies*, Vol. 19, No. 2.

EXERTZOGLOU, H. (2000), "H Sigkrotisi tou Dimosiou Horou stin Konstantinoupoli ton 19o aiona" (The Formation of Public Space in Constantinople in the 19th Century), in *Greeks Outside, Constantinople and Smyrna 1800-1922: intellectual and social life*. Symposium October 1998, Athens, Etaireia Spoudon Neoellinkou Politismou kai Genikis Paideias.

FANON, F. (1952), *Black Skin, White Masks*, New York, Grove Press.

FARGION, V. (2005), "From the Southern to the Northern Question", in McEWEN, N. McEwen and MORENO, L. (eds.), *The Territorial Politics of Welfare*, London, Routledge.

FEDI, M. (2008), Electoral Speach 2008 per la Camera – per il bene dell'Italia e degli italiani all'estero, http://www.youtube.com/watch?v=2XTLTFt0Y.

FEDI, M. (2009), marcofedi.blogspot.com.

FERRERA, M. (2005), *The Boundaries of Welfare. European Integration and the New Spatial Politics of Social Protection*, Oxford, Oxford University Press.

FINNANE, M. (2003), "From dangerous lunatic to human rights? The law and mental illness in Australian history", in COLERBONE, C. and MCKINNON, D. (eds.), *Madness in Australia*, St. Lucia, University of Queensland Press.

FISHER, F.G. (1994), *Raphael Cilento. A Biography*, St. Lucia, University of Queensland Press.

FIVE DOCK RSL COMMUNITY CLUB, Club Five Dock, www.pc.gov.au/__data/assets/pdf_file/0005/87584/sub141.pdf.

FONTANA, S. M. C. (No date), *A Story Is Told: Internment – Truth, Memory and Reconciliation*, Bunbury, W.A., Harvey Shire Council.

FORRESTER, D. (1991), "The Depressed Classes and Conversion to Christianity 1860-1960", in ODDIE, G. (ed.), *Religion in South Asia: Religious Conversion and Revival Movements in South Asia in Medieval and Modern Times*, 2nd ed., New Delhi, Manohar Books.

FORZA ITALIA (2003), "Documento Programmatico", in *Azzurri nel mondo*, http://www.azzurrinelmondo.it/images/Documento%20programmatico.pdf.

FOUCAULT, M. (1980), *Power/Knowledge*, New York, Pantheon Books.

FOUCAULT, M. (1986), "Of Other Spaces", *Diacritics*, Vol. 16, No. 1.

FOUGÈRES, G. (1918), *H Apostoli tou Ellinismou kai to Kathikon tou Ellinos Stratiotou* (The Mission of Hellenism and the Greek Soldier's Duty) (FOTIADIS, A., trans. from French), Athens, Typografeion P. A. Petrakou.

FRANKENBERG, R. (1997), *Displacing Whiteness. Essay in Social and Cultural Criticism*, Durham and London, Duke University Press.

FREDRICKSON, G.M. (1971), *The Black Image in the White Mind*, Middletown (Conn.), Wesleyan University Press, 1987.

FRIGESSI, D. (2003), *Cesare Lombroso*, Torino, Einaudi.

GABACCIA, D. (2000), *Italy's Many Diasporas*, London, UCL Press.

GALLANT, T. (1997), "Greek Exceptionalism and Contemporary Historiography: New Pitfalls and Old Debates", *Journal of Modern Greek Studies*, Vol. 15, No. 2.

GAZZINI, C. (2009), "Assessing Italy's *Grand Gesto* to Lybia", Middle East Report Online, March 16, http://www.merip.org/mero/mero031609.html.

GHORASHI, H. (2003), "Ayaan Hirsi Ali: Daring or Dogmatic. Debates on multiculturalism and emancipation in the Netherlands", *Focaal*, Vol. 42. No. 163.

GHORASHI, H. (2005), "Benauwd door de Verlichting. De 'Witheid' van de Huidige Debatten over Emancipatie", *Eutopia*, No. 10.

GIACANELLI, F. (1975), *Introduzione*, in COLOMBO, G. (ed.), *La scienza infelice. Il museo di antropologia criminale di Cesare Lombroso*, Torino, Boringhieri.

GIANNACOPOULOS, M. (2007), "Mabo, Tampa and the Non-Justiciablity of Sovereignty", in PERERA, S. (ed.), *Our Patch: Enacting Australian Sovereignty Post 2001*, Perth, Network Books.

GIBBS, A. (2001), "Contagious Feelings: Pauline Hanson and the Epidemiology of Affect", in *Australian Humanities Review*, http://www.lib.latrobe.edu.au/AHR/archive/Issue-December-2001/gibbs.html.

GIBSON, M. (2002), *Born to Crime: Cesare Lombroso and the Origins of Biological Criminology*, Westport, Praeger.

GILCRHRIST, H. (1997), *Australians and Greeks. Volume II: The Middle Years*, Sydney, Halstead Press.

GILROY, P. (2000), *Against Race: Imaging Political Culture Beyond the Color Line*. Cambridge, MA, The Belknap Press of Harvard University Press.

GILROY, P. (1993), *The Black Atlantic: Modernity and Double Consciousness*. London, Verso.

GIULIANI, G. (2007), "Fascismo e *bio*politica. L'editoria emiliano-romagnola e la questione della razza", in TORTORELLI, G. (ed.), *Editoria e cultura in Emilia-Romagna dal 1900 al 1945*, Bologna, Compositori.

GIULIANI, G. (2008), *Beyond curiosity. James Mill e la nascita del governo coloniale britannico in India*, Roma, Aracne.

GIULIANI, G. (forthcoming), 'Selecting Whites: Fantasie di Bianchezza nell' Australia Indipendente,' *Studi Culturali*, Vol. 1.

GRAMSCI, A. (1971), *Selections from the Prison Notebooks of Antonio Gramsci*, trans. and ed., Q. Hoare, G. N. Smith, New York, International Publishers.

GRAMSCI, A. (1975), *Quaderni del Carcere*, Vol. 3, Torino, Giulio Einaudi.

GRAMSCI, A. (2005), *La Questione Meridionale*, Roma, Editori Riuniti.

GREWAL, I. 2005, *Transnational America: Feminisms, Diasporas, Neoliberalism*, Durham and London, Duke University Press.

GUILLAUMIN, C. (1972), "Caractères spécifiques de l'idéologie raciste", *Cahiers Internationaux de sociologie*, Vol. LIII.

GUNEW, S. (1994), *Framing Marginality: Multicultural Literary Studies*, Carlton, Vic., Melbourne University Press.

HAGE, G. (1998), *White Nation: Fantasies of White Supremacy in a Multicultural Society*, Sydney, Pluto Press.

HAGE, G. (2004), "L'estetica di classe nel multiculturalismo australiano", in MEZZADRA, S. (ed.), *I confini della libertà. Per un'analisi politica delle migrazioni contemporanee*, Roma, Derive Approdi.

HALL, S. (1996), "New Ethnicities", in MORLEY, D. and CHEN, (eds.), *Stuart Hall: Critical Dialogues in Cultural Studies*, London, Routledge.

HARLOW, B. (1986), "Return to Haifa: 'Open the Borders' in Palestinian Literature", *Social Text*, No. 13/14.

HEKIMOGLOU, E. (1997), "Thessaloniki, 1912-1940: Economic Developments", in CHASSIOTIS, I. K. (ed.), *Queen of the worthy: Thessaloniki, history and culture*, Vol. 1, Thessaloniki, Paratiritis.

HELSTOSKY, C. (2004), *Garlic and Oil: Food and Politics in Italy*, Oxford and New York, Berg.

HERALD SUN. (2008), "Italian sunbathers ignored drowned Gypsy girls, 22 July 2008", http://www.news.com.au/heraldsun/story/0,21985,24059338-661,00.html.

HERZFELD, M. (1986), *Ours Once More: Folklore, Ideology, and the Making of Modern Greece*, New York, Pella Publishing Company.

HINDUSTAN TIMES. (2007), "Protest Planned Against Sonia's UN Visit", 9 September. http://www.hindustantimes.com.

HIRSCHON, R. (ed.) (2003), *Crossing the Aegean: An Appraisal of the 1923 Compulsory Population Exchange Between Greece and Turkey*, New York, Berghahn Books.

HIRSI ALI, A. (2002), "Schurende normen. Integratie als inwijding in de moderniteit", in F. Becker, W. van Hennekeler, M. Sie Dhian Ho & B. Tromp (eds.), *Transnationaal Nederland. Het drieëntwintigste jaarboek voor het democratisch socialisme*, Amsterdam, WBS/Arbeiderspers.

HIRSI, ALI, A. (2002b), "Integratie is een cultureel probleem", *NRC Handelsblad*, 31 August.

HIRSI, ALI, A. (2003), *De zoontjesfabriek*, Amesterdam and Antrwerpen, Pubisher Augustus.

HIRSI, ALI, A. (2004), *De maagdenkooi*, Amesterdam and Antwerpen, Publisher Augustus.

HOLT, P.M., and Daly, M.W. (1998), *A History of the Sudan: From the Coming of Islam to the Present Day*, London, Longman.

HULME, P. (2005), "Beyond the Straits: Postcolonial Allegories of the Globe", in LOOMBA A., KAUL, S., M. BUNZL, M., BURTON, A., and ESTY, J. (eds.), *Postcolonial Studies and Beyond*, Durham, Duke University Press.

HUNWICK, J. O. and POWELL, E. T. (2002), *The African Diaspora in the Mediterranean Lands of Islam*, Princeton, Markus Wiener Publishers.

HUSSEIN, A. (2002), *Edward Said: Criticism and Society*, London, Verso.

IACAVINO, V. (1983), *Legislative and Administrative Requirements in Relation to Italian Migration to Australia During the Inter-war Years*, Carlton, Victoria, Co.As.It, Italian Historical Society.

INDO-ASIAN NEWS SERVICE. (2008), "VHP continues attack against Sonia Gandhi", 25 November, http://www.indiaforums.com/news/article.asp?id=138488.

IZZIO, F. (1999), *I Lager dei Savoia: Storia infame del Risorgimento nei campi di concentramento per meridionali*, Napoli, Controcorrente.

JACOBS, J.M. (1996), *The Edge of Empire. Postcolonialism and the City*, London and New York, Routledge.

JACOBSON, M.F. (1998), *Whiteness of a Different Colour, European Immigrants and the Alchemy of Race*, Harvard, Harvard University Press.

JAFFRELOT, C. (1999), *The Hindu Nationalist Movement and Indian Politics, 1925 to the 1990s*, New Delhi, Viking.

JORDAN, W. (1977), *White Over Black*, London, W.W. Norton & Company.

JORDENS, A-M. (1997), *Alien to Citizen: Setting Migrants in Australia*, 1945-75, Sydney, Allen and Unwin.

KAFKA, F. (1992), *The Transformation ('Metamorphosis') and Other Stories*, trans. PASLEY, M., London, Penguin.

KAMAT, S. and BIJU, M. (2003), "Mapping Political Violence in a Globalized World: The Case of Hindu Nationalism", *Social Justice* Vol. 30, No. 3.

KANARAKIS, G. (1997), *In the Wake of Odysseus: Portraits of Greek Settlers in Australia*, Melbourne, RMIT University Press.

KARAMBLIAS, I. (1949), *Istoria ton Kidonion: Apo tis idriseos ton mehri tis apokatastaseos ton prosfigon eis to eleftheron Ellinikon Kratos* (History of Kydonies: From its founding till the resettlement of the refugees in the free Greek State), Vols. 1 & 2, Athens, Tipografeion Ger. S. Christou & Yios.

KATSAPIS, K. (2003a), "Antiparatheseis anamesa se gigeneis kai prosfiges stin Ellada tou Mesopolemou" (Conflicts between native and refugee in Greece in the interwar years), in TZEDPOULOS, G. (ed.), *Beyond the Catastrophe: Asia Minor Greeks in Interwar Greece*, Athens, Idrima Meizonos Ellinismou, Ipo tin Aigida tis Voulis ton Ellinon.

KECHROTIS, V. (2005), "Greek-Orthodox, Ottoman Greeks or just Greeks? Theories of Coexistence in the Aftermath of the Young Turk Revolution", *Études Balkaniques*, Vol. 1.

KIRK-GREENE, A. H. M. (1980), "The Thin White Line: The Size of the British Colonial Service in Africa", *African Affairs*, Vol. 79, No. 314.

KITROEFF, A. (1989), *The Greeks in Egypt 1919-1937: Ethnicity and class*, London, Ithaca Press for The Middle East Centre, St Antony's College Oxford.

KITROEFF, A. (1991), "The Transformation of the Homeland-Diaspora Relations: The Greek Case in the 19th-20th centuries", in FOSSEY, J. M. (ed.), *Proceedings of the First International Congress on the Hellenic Diaspora from Antiquity to Modern Times*, Vol. 2, Amsterdam, Gieben.

KITROEFF, A. (2002), "Emporikes paroikies kai metanastes" (Trading Communities and Migrants), in CHATZEIOSEPH, C. (ed.), *The History of Greece in the 20th Century: 1922-1940 The Interwar*, Vol. 2 (Part 1), Athens, Vivliorama.

KITROMILIDES, P. (2003), "In the Pre-Modern Balkans... Loyalties, Identities, Anachronisms", in TZIOVAS, D. (ed.), *Greece and the Balkans: identities, perceptions and cultural encounters since the Enlightenment*, Aldershot, Ashgate.

KONTOGIORGI, E. (1992), "Agrotikes Prosfigikes Egkatastaseis sti Makedonia 1923-1930" (Rural Refugee Settlements in Macedonia 1923-1930), *Deltio Kentrou Mikrasiatikvon Spoudon* (Bulletin of the Centre for Asia Minor Studies), Vol. 9.

KONTOGIORGI, E. (2003), "Economic Consequences following Refugee Settlement in Greek Macedonia, 1923-1932", in HIRSCHON, R. (ed.), *Crossing the Aegean: An Appraisal of the 1923 Compulsory Population Exchange Between Greece and Turkey*, New York, Berghahn Books.

KOSIC, A. and TRIANDAFYLLIDO, A. (2005), "Active Civic Participation of Immigrants in Italy", Country Report for European Research Project POLITIS, Oldenburg, www.uni-oldenburg.de/politis-europe.

KRAUT, R. (1984), *Socrates and the State*, Princeton, New Jersey, Princeton University Press.

LA REPUBBLICA.IT (2001), "Recuperate i corpi della nave fantasma", 4 Agosto 2001, http://ricerca.repubblica/archivio/repubblica/2001/08/04/recuperate-corpi-della-nave-fantasma.html.

LA REPUBBLICA.IT. (2002), "Portopalo, fiori e preghiere tamil per i morti del naugragio fantasma", 8 Aprile 2002, http://ricerca.repubblica.it/repubblica/archivio/repubblica/2002/04/08/portopalo-fiori-preghiere-tamil-per-morti-de.html.

LALIOTOU, I. (2004), *Transatlantic Subjects: Acts of Migration and Cultures of Transnationalism between Greece and America*, Chicago, University of Chicago Press.

LANDUCCI, F. (1972), *I filosofi e i selvaggi*, Bari, Laterza.

LANGFORD GINIBI, R. (1994), *My Bundjalung People*, St Lucia, University of Queensland Press.

LANGFORD, S. (1997), "Appendix", in EDWARDS, P. (ed.), *A Nation at War: Australian Politics, Society and Diplomacy during the Vietnam War 1965-1975*, Sydney, Allen and Unwin.

LAPIDUS, I. M. (2002), *A History of Islamic Societies*, Cambridge, Cambridge University Press.

LAZARUS, N. (1997), 'Transnationalism and the alleged death of the nation state,' in ANSELL-PEARSON, K., PARRY, B. and SQUIRES, J. (eds.), *Cultural Readings of Imperialism: Edward Said and the Gravity of History*, New York, St. Martin's Press.

LEAGUE OF NATIONS (1926), *Greek Refugee Settlement,* Geneva, Publications of the League of Nations.

LEE, B. and LIPUMA, E. (2002), "Cultures of Circulation: The Imaginations of Modernity". *Public Culture*, Vol. 14, No. 1.

LEED, E. J. (1979), *No man's land: combat & identity in First World War*, Cambridge, Cambridge University Press.

LEWIS, B. (1992), *Race and Slavery in the Middle East: An Historical Enquiry*, Oxford, Oxford University Press.

LIAKOS, A. (1988), *I Emfanisi ton Neanikon Organoseon: to paradeigma tis Thessalonikis* (The Emergence of Youth Organisations: the example of Thessaloniki), Athens, Lotos.

LIAKOS, A. (1998, 13 September), "I Ideologia ton 'Hamenon Patridon'" (The Ideology of the 'Lost Homelands'), *To Vima*, B01 Internet Archival Code Number: B12498B12011 ID: 108296.

LIAKOS, A. (2002), "The Construction of National Time: The Making of the Modern Greek Historical Imagination", in REVEL, J. and LEVI, G. (eds.), *The Political Uses of the Past: The Recent Mediterranean Experience*, London and Portland, OR., Frank Cass.

LIENHARDT. G. (1982), "The Sudan: Aspects of the South Government among Some of the Nilotic Peoples, 1947-52", *Bulletin (British Society for Middle Eastern Studies)*, Vol. 9, No. 1.

LIPSITZ, G. (1998), *The Possessive Investment in Whiteness. How White People Profit from Identity Politics*, Philadelphia, Temple University press.

LOCKMAN, Z. (2004), *Contending Visions of the Middle East: The History and Politics of Orientalism*, Cambridge, Cambridge University Press.

LOMBOROSO, C. (1868), *Azione degli astri e delle meteore sulla mente umana ed alienata*, Milano, Raedelli.

LOMBROSO, C. (1871), *L'uomo bianco e l'uomo di colore. Letture su l'origine e la varietà delle razze umane*, Torino, Bocca, 1892b, 2nd ed.

LOMBOROSO, C. (1876), *L'uomo delinquente studiato in rapporto alla antropologia, alla medicina legale ed alle discipline carcerarie*, Milano, Ulrico-Hoepli.

LOMBROSO, C. (1878), *Pensiero e meteore. Studi di un alienista*, Milano, F.lli Dumolard.

LOMBROSO, C. (1892), G. Ferrero, *La donna delinquente. La prostituta e la donna normale*, Torino.

LOMBOROSO, C. (1894), *Gli anarchici*, Roma, Napoleone.

LOMBROSO, C. (1895), *L'Homme Criminel*, Felix, Alcan.

LOMBROSO, C. (1898 [1862]), *In Calabria (1862-1897)*, Catania, Cav. Niccolò Giannotta.

LOMBROSO, C. (1911), *Criminal Man, According to the Classification of Cesare Lombroso*, with Gina Lombroso-Ferrero, New York, Putnam.

LOMBROSO, C. (2006), *Criminal Man*, GIBSON, M. and RAFTER, N. H. (trans), Durham, Duke University Press.

LONDON COMMITTEE OF UNREDEEMED GREEKS (1919a), *Short account of the Turkish persecutions and atrocities committed against the Greeks in Turkey*, London, London Committee of Unredeemed Greeks.

LONDON COMMITTEE OF UNREDEEMED GREEKS (1919b), *The Fall of the Empire of the Sultans and the rights of Hellenism*, London, The London Committee Of Unredeemed Greeks.

LOPEZ, M. (2000), *The Origins of Multiculturalism in Australian Politics 1945-1975*, Melbourne, Melbourne University Press.

LOUIS, W. R. (2006), *Ends of British imperialism: the scramble for empire, Suez and decolonization: collected essays*, London, I.B.Tauris.

LYNG, J. (1935), *Non-Britishers in Australia; Influence on Population and Progress*, Melbourne, Melbourne University Press.

MABO AND OTHERS v STATE OF QUEENSLAND (No. 2) (1991), 107 ALR1, http://www.austlii.edu.au/au/cases/cth/HCA/1992/23.html.

MACKRIDGE, P. (2003), "The Myth of Asia Minor in Greek Fiction", in HIRSCHON, R. (ed.), *Crossing the Aegean: An Appraisal of the 1923 Compulsory Population Exchange Between Greece and Turkey*, New York, Berghahn Books.

MACKRIDGE, P. and YANNAKIS, E. (eds.), (1997) *Ourselves and Others. The Development of a Greek Macedonian Cultural Identity Since 1912*, Oxford New York, Berg.

MACMICHAEL, H.A. (1967), *A History of the Arabs in the Sudan: and some account of the people that preceded them and of the tribes of Darfur*, London, Frank Cass.

MAHMOOD, S. (2001), "Feminist Theory, Embodiment, and the Docile Agent: Some Reflections on Egyptian Islamic Revival", *Cultural Anthropology*, Vol. 16, No. 2.

MAMDANI, M. (2009), *Saviors and Survivors: Darfur, Politics, and the War on Terror*, New York, Pantheon Books.

MANNE, R. (2007), "A Turkish Tale. Gallipoli and the Armenian Genocide", *Monthly*, No. 20 (February), www.themonthly.com.au/tm/node/459.

MAPS OF WORLD.COM (2009), "La Maddalena and Lampedusa, Italy", http://www.mapsofworld.com/italy/italy-islands/la-maddalena-and-lampedusa-italy.html.

MARCHAND, M. P. J. (eds.) (1995), *Feminism/Postmodernism/Development*, New York, Routledge.

MARISCAL, J. (2003), "Military Targets Latinos: Tracked into Combat Jobs", in *Wartimes*, Issue 13, http://www.war-times.org/issues/13art5.html.

MARKUS, A. (1994), *Australian Race Relations 1788-1993*, St. Leonards, Allen & Unwin.

MARTUCCI, R. (2007), *L'Invenzione dell' Italia Unita*, Milano, Sansoni.

MASSAD, J. (2004), "The Intellectual Life of Edward Said", *Journal of Palestinian Studies*, University of California Press.

MASSAD, J. (2008), "Beginning with Edward Said", in SOKMEN, M. and ERTUR, B. (eds.), *Waiting for the Barbarians*, London, Verso.

MAZOWER, M. (2004), *Salonica, City of Ghosts: Christians, Muslims and Jews, 1430-1950*, New York, Alfred A. Knopf.

MAZRUI, A. A. (2005), "The Re-invention of Africa: Edward Said, V. Y. Mudimbe, and Beyond", *Research in African Literatures*, Vol. 36, No. 3.

MBEMBE, A. (2003), "Necropolitics", *Public Culture*, Vol. 15, No. 1.

McCLINTOCK, A. (1995), *Imperial Leather: Race, Gender and Sexuality in the Colonial Conquest*, London, Routledge.

MOHANTY, C. (1991), *Third World Women And The Politics of Feminism*, Bloomington, Indiana University Press.

MOHNRAM, R. (2007), *Imperial White. Race, Diaspora, and the British Empire*, Minneapolis, University of Minnesota Press.

MOORS, A., TEEFELEN, T., KANAAN, S. and ABU GHAZALEH, I. (eds.) (1995), "Exile, Diaspora, and Return: the Inscription of Palestine in Zionist Discourse", *Discourse and Palestine: Power, Text and Context*, Amsterdam, Het Spinhuis.

MORETON-ROBINSON, A. (2003), "I still call Australia home: Indigenous Belonging and Place in a White Postcolonising Society", in AHMED, S., CASTANEDA, C. and A.M. FORTIE (eds.), *Uprootings/Regroundings: Questions of Home and Migration*, Oxford, Berg Publishing.

MORETON-ROBINSON, A. (2004), "The Possessive Logic of Patriarchal White Sovereignty: The High Court and the Yorta Yorta Decision", *Borderlands ejournal* 3.2, http://www.borderlandsejournal.adelaide.edu.au/vol3 no2_2004/moreton_possessive.htm.

MORETON-ROBINSON, A. (2005), "Whiteness, Epistemology and Indigenous Representation", in MORETON-ROBINSON, A. (ed.), *Whitening Race: Essays in Social and Cultural Criticism*, Canberra, Aboriginal Studies Press.

MOSKOF, K. (1982), "To Aïvali – ideologia enos horou" (Aivali – the ideology of a place), in SAKKARIS, G. (ed.), *History of Kydonies*, Athens, Tipois Al. Vitsikounakis (Anatiposi apo Ekdosi Enosis Kidoniaton).

MOULLAS, P. (1993), "Eisagogi stin Mesopolemiki Pezografia" (Introduction to Interwar Prose), MOULLAS, P. (ed.), *Interwar Prose: From the First to the Second World War*, Vol. 1, Athens, Ekdoseis Sokoli.

MUDIMBE, V.Y. (1988), *The Invention of Africa: Gnosis, Philosophy, and the Order of Knowledge*, Bloomington, University of Indiana Press.

NANDY, A. (1998), "The twilight of certitudes: secularism, Hindu nationalism and other masks of Deculturation", *Postcolonial Studies*, Vol. 1, No. 3.

NATIONAL ARCHIVES OF AUSTRALIA *Fact Sheet 164-National Service 1965-72*, http://www.naa.gov.au/about-us/publications/fact-sheets/fs164.aspx.

NEILSON, B. (2005), "La politica dell'immaginario. Appunti incompleti su affetti e potere", *Studi Culturali*.

NINIAN, A. (2004), "The Indian General Election of 2004", *Contemporary Review*, 285, 1662.

O'FAHEY, R.S. (1997), "Endowment, privilege and estate in the Central and Eastern Sudan'", *Islamic Law and Society*, Vol. IV, No. 3.

O'FAHEY, R.S. (1980), *State and Society in Darfur*, London, Hurst and Co.

O'FAHEY, R. (2006), "Does Darfur have a future in the Sudan?", *The Fletcher Forum of World Affairs*, Vol. 30, No. 1.

O'FAHEY, R.S. (2008), *The Darfur Sultanate*, London, Hurst and Co.

O'HANLON, R. (1994), *A Comparison Between Women and Men: Tarabai Shinde and the Critique of Gender Relations in Colonial India*, New Delhi, Oxford University Press.

ONG, A. (1988), "The Production of Possession: Spirits and the Multinational Corporation in Malaysia", *American Ethnologist*, Vol. 15, No. 1.

ONG, A. (2006), *Neoliberalism as Exception: Mutations of Citizenship and Sovereignty*, Durham, North Carolina, Duke University Press.

OSURI, G. and BANERJEE, SB. (2004), "White Diasporas: Media Representations of September 11 and the Unbearable Whiteness of Being in Australia", *Social Semiotics*, Vol. 14, No. 2.

PAGANO, A. (2002), "Breve Storia delle Due Sicilie", http://www.duesicilie.org/spip.php?article48&var_recherche=Pagano.

PALANO, D. (2002), *Il potere della moltitudine. L'invenzione dell'inconscio collettivo nella teoria politica e nelle scienze sociali italiane tra Otto e Novecento*, Milano, Vita e Pensiero.

PALOMBO, L. (2007), "Whose Turn is it? White Diasporic and Transnational Practices and the Necropolitics of the Plantation and Internment Camps", *ACRAWSA e-Journal*, Vol. 3, No. 7.

PANICHI, J. (2008), "Diario Elettorale", *La Fiamma*, 21 March.

PAPADOPOULOS, A. (1927), *International Directory of 1927 – O Diethnis Emporikos Odigos*, Adelaide, International Publishers.

PAPADOPOULOS, D. and TSIANOS, V. (2007), "Introduction", in RAJARAM, P. and GRUNDY-WARR, C. (eds.), *Borderscapes: Hidden geographies and politics of territory's edge*, Minneapolis, University of Minnesota Press.

PAPAILIAS, P. (2005), *Genres of Recollection: Archival Poetics and Modern Greece*, New York, Palgrave Macmillan.

PAPANDREOU, A. (1973), *Democracy at Gunpoint: The Greek Front*, Harmondsworth, Penguin.

PARKER, I. (1992), *Discourse Dynamics: Critical analysis for social and individual psychology*, London and New York, Routledge.

PARLIAMENT OF THE COMMONWEALTH OF AUSTRALIA (1966), House of Representatives, No. 174, 31st August.

PARPART, J. (2000), "The Participatory Empowerment Approach to Gender and Development in Africa: Panacea or Illusion?", in PARPART, J., RAI, S. and STAUDT, K. (eds.), *Rethinking Empowerment in a Global World*, New York, Routledge.

PARTITO DEMOCRATICO, (2008a), Partito Democratico in Australia, http://pdaustralia.blogspot.com/2008/02/partito-democratico-in-australia.html.

PARTITO DEMOCRATICO, (2008b), Partito Democratico in Australia, http://pdaustralia.blogspot.com/2008/03/italians-abroad-resource.html.

PARTITO DEMOCRATICO, PD (2008), "Standing for the Italian Parliament", Political Brochure.

PARTRIARCA, S. (2005), "Indolence and Regeneration: Tropes and Tensions of Risorgimento Patriotism", *The American Historical Review*, Vol. 110, No. 2, http://www.historycooperative.org/journals/ahr/110.2/patriarca.html.

PDL AUSTRALIA (2008), Pubblicità dei Candidati Teresa Todaro Restifa e Luigi Casagrande per il Popolo della Libertà in Africa Asia Oceania e Antartide, http://www.youtube.com/watch?v=cJ6lxfHtYHk.

PENDE, N. (1930), "Psicologia individuale e psicologia di razza", *Rivista di psicologia*, Vol. XXVI.

PENDE, N. (1933), *Bonifica umana razionale e biologia politica*, Bologna, Cappelli.

PERERA, S. (2002), "What is a Camp...?", *Borderlands*, Vol. 1, No. 1, http://www.borderlands.net.au/vol1no1_2002/perera_camp.html.

PERERA, S. (2002a), "A line in the sea: the *Tampa*, boat stories and the border", Cultural *Studies Review*, Vol. 8, No. 1.

PERERA, S. (2006), "'They Give Evidence': Bodies, Borders and the Disappeared", Social *Identities*, Vol. 12, No. 6.

PERERA, S. (2007), "Our Patch: Domains of Whiteness, Geographies of Lack and Australia's Racial Horizons in the War on Terror", in PERERA, S. (ed.) *Our Patch. Enacting Australian Sovereignty Post 2001*, Perth, Network Books.

PERERA, S. (2008), "A Pacific Zone? (In)Security, Sovereignty, and Stories of the Pacific Borderscape", in RAJARAM, P.K. AND GRUNDY-WARR, C. (eds.), *Borderscapes: Hidden Geographies and Politics at Territory's Edge*, Minneapolis, University of Minnesota Press.

PERERA, S. (2009), *Australia and the Insular Imagination: Borders, Beaches, Boats and Bodies*, New York, Palgrave.

PETMEZAS, S. (2002), "Agrotiki Oikonomia" (Agricultural Economy). In CHATZEIOSEPH, C. (ed.), *The History of Greece in the 20th Century: 1922-1940 The Interwar*, Vol. 2, Part 1, Athens, Vivliorama.

PETRACCONE, C. (2000), *Le Due Civiltà*, Roma and Bari, Laterza.

PETRIDIS, P. (1997), "Thessaloniki, 1912-1940: A Period of Political and Social Transition", in CHASSIOTIS, I. K. (ed.), *Queen of the worthy: Thessaloniki, history and culture*, Vol. 1, Thessaloniki, Paratiritis.

PETROPOULOU, I. (1982), "Giro apo tin Istoriografia ton Kidonion. Bibliokritiko Dokimio gia tis I. Kerestetzi; *Aïvali 1832-1922* Athina 1981" (Regarding the Historiography of Kydonies. A Review Essay of I. Kerestetzi *Aivali 1832-1922* Athens 1981), *Deltio Kentrou Mikrasiatikon Spoudon* (Bulletin of the Centre for Asia Minor Studies), Vol. 3.

PETTMAN, J. J. (1992), *Living in the Margins: Racism, Sexism and Feminism in Australia*, Sydney, Allen and Unwin.

PIPER, K. (2002), *Cartographic Fictions: Maps, Race, and Identity*, New Brunswick, NJ, Rutgers University Press.

PLATO (1954), *The Last Days of Socrates: Euthyphro, The Apology, Crito, Phaedo*, n.p., Penguin.

POLIAKOV, L. (1971), *The Aryan Myth. A History of Racist and Nationalistic Ideas In Europe*, New York, Barnes & Noble Books.

POLLICINA, C. (2008), "Tra noi", *La Fiamma*, 21 March.

POLYBUS (1919), *The Settlement of Turkey*, London, East and West Ltd., reprinted from the *Asiatic Review*, January 1919.

PORTES, A. (1999), "Conclusion: Towards a New World – The Origins and Effects of Transnational Activities", *Ethnic and Racial Studies*, Vol. 2, No. 2.

PÖTZSCH, H. "Tracing Topographies of the Exception: Borders, Zombies and Bare Life", http://www.ntnu.no/eksternweb/multimedia/archive/00043/ HolgerP_tzsch_43229a.pdf.

PROGETTO MELTING POT EUROPA. (2009), "Lampedusa – There are serious and imminent risk of widespread violations of the fundamental right of refugees and immigrants", 27 January, http://www.meltingpot.org/ stampa13895.html.

PUGLIESE, J. (1995), "Assimilation, Unspeakable Traces and the Ontologies of Nation", in PERERA, S. (ed.), *Asian and Pacific Inscriptions. Identities, Ethnicities, Nationalities*, Melbourne, Meridian.

PUGLIESE, J. (2002), "Race as Category Crisis: Whiteness and the Topical Assignation of Race", *Social Semiotics*, Vol. 12, No. 2.

PUGLIESE, J. (2002a), "In the Ruins of Diaspora", in SIEGEL, K. (ed.), *Issues in Travel Writing: Empire, Spectacle, and Displacement*, New York, Peter Lang.

PUGLIESE, J. (2006), "Bodies of Water", *HEAT*, Vol. 12.

PUGLIESE, J. (2007), "White Historicide and the Returns of the Souths of the South", *Australian Humanities Review*, Vol. 42, http://www.lib.latrobe. edu.au/AHR/archive/Issue-August-2007/Pugliese.html.

PUGLIESE, J. (2007a), "Geocorpographies of Torture", *Australian Critical Race and Whiteness Studies ejournal*, Vol. 3, No. 1, http://www.acrawsa.org. au/ejournalFiles/Volume%203,%20Number%201,%202007/JosephPugliese.pdf

PUGLIESE, J. (2008), "Whiteness and the Blackening of Italy: *La Guerra Cafona, Extracommunitari* and Provisional Street Justice", *Portal: Journal of Multidisciplinary International Studies*, Vol. 5, No. 2.

PUGLIESE, J. (2009), "Civil Modalities of Refugee Trauma, Death and Necrological Transport", *Social Identities*, Vol. 15, No. 1.

RAJARAM, P. and GRUNDY-WARR, C. (2007), "Introduction", in RAJARM, P. and GRUNDY-WARR, C. (eds.) *Borderscapes: Hidden geographies and politics of territory's edge*, Minneapolis, University of Minnesota Press.

RALLI, V. (1995), *Patrida Aksehasti Mikra Asia: i tragodia tou Mikrasiatikou Ellinismou mesa apo tin istoria mias Moschonisiotikis oikogeneias* (Unforgettable Country. Asia Minor: The Tragedy of Asia Minor Greeks through the Story of a Moschonisiot Family), Athens, Ekdoseis Akritas.

RANDAZZO, N. (2008), Election Video 2008, per il Senato – per il bene dell'Italia e degli italiani all'estero, http://www.youtube.com/watch?v=td FqmrMqBwU.

RANGER. T. (1992), "The Invention of Tradition in Colonial Africa", in HOBSBAWN, E. and RANGER, T. (eds.), *The Invention of Tradition*, Cambridge, Cambridge University Press.

RAPPORTO CONGUINTO OPEN SOCIETY (2009), http://www.unione dirittiunmani.it.

REICHMUTH, S. (2004), "Murtada al Zabidi (1732-91) and the Africans: Islamic discourse and the scholarly networks in the late 18th century", in REESE, S.S. (ed.), *The transmission of learning in Islamic Africa*, Leiden, Koninklijke Brill NV.

RESSA, G. (2008), "Il Sud e L'Unità D'Italia", http://www.cittanovattiva.it/ public/il_sud_e_l_unità_d_Italia.pdf.

REYNOLDS, H. (2005), *Nowhere People. How International Race Thinking Shaped Australia's Identity*, Victoria, Viking & Penguin.

RIFUGIATI DI MILAN (2009), rifugiatimilano.blogspot.com/2009/04/ cronistoriasempreaggiornatadelle.html.

ROSSER, B. (1978), *This is Palm Island*, Canberra, Australian Institute of Aboriginal Studies.

ROUX, C. (2008), "Italy's Path to Federalism. Origins and Paradoxes", *Journal of Modern Italian Studies*, Vol. 13, No. 3.

RUZZA, C. and SCHMIDTKE, O. (2007), "Changing Conceptions of the Enemy in the Lega Lombarda", in RAJARAM, P. and GRUNDY-WARR, C. (eds.), *Borderscapes: Hidden geographies and politics of territory's edge*, Minneapolis, University of Minnesota Press.

SAFRAN, W. (1991), "Diasporas in Modern Societies: Myths of Homeland and Return", *Diaspora: a journal of transnational studies*, Vol. 1, No. 1.

SAID, E. ([1978] 1985), *Orientalism: Western Conceptions of the Orient, London*, Penguin.

SAID, E. (1979), 'Zionism from the Stand Points of Its Victims', *Social Text*, No. 1.

SAID, E. (1983), *The World, the Text and the Critic*, Cambridge, MA, Harvard University Press.

SAID, E. (1985), *Beginnings: intention and method*, New York, Columbia University Press.

SAID, E. (1993), *Culture and Imperialism*, New York and London, Vintage Books.

SAID, E. (1994), *Culture and Imperialism*, London, Vintage.

SAID, E. (2000), *Reflection on Exile and Other Essays*, Cambridge, Harvard University Press.

SAID, E. (2000), "The Claims of Individuality' (1966)", in BAYOUMI, M. and RUBIN, A. (eds.), *The Edward Said Reader*, London, Granta Books.

SAID, E. (2000), 'The Palestinian Experience' (1968-1969), in BAYOUMI, M. and RUBIN, A. (eds.), *The Edward Said Reader*, London, Granta Books.

SAID, E. (2001), "My Guru", *London Review of Books*.

SAKKARIS, G. ([1920] 1982), *Istoria ton Kidonion* (History of Kydonies), Athens, Tipois Al. Vitsikounakis (Reprint by Ekdosi Enosis Kidoniaton).

SALERA, L. (2006), *Garibaldi, Fauché e i Predatori del Regno del Sud*, Napoli, Controcorrente.

SALERNO, E. (1979), *Genocidio in Libia: Le Atrocita Nascoste Dell'Avventura Coloniale (1911-1931)*, Milano, SugarCo.

SARKAR, S. (1999), "Hindutva and the Question of Conversions", in PANIKKAR, K. N. (ed.), *The Concerned Indian's Guide to Communalism*, New Delhi, Viking Penguin.

SARKAR, T. (2001), *Hindu Wife, Hindu Nation: Community, Religion, and Cultural Nationalism*, Bloomington and Indianapolis, Indiana University Press.

SARTI, R. (1997), *Mazzini: a life for the religion of politics*. Santa Barbara, CA, Greenwood Publishing.

SCHNEIDER, J. (ed.) 1998, *Italy's "Southern Question": Orientalism in One Country*, Oxford-New York, Berg.

SELTIN, J. (2008), "Upcoming Symposium: Relational Dis/Locations: Mediterranean Cultures in Translocal and Transnational Contexts", November 25, Email Communication.

SEN, A. (1999), *Development as Freedom*, Oxford, Oxford University Press.

SHARKEY, H. J. (2003), *Living with Colonialism: Nationalism, and Culture in the Anglo-Egyptian Sudan*, Berkeley, University of California Press.

SHARKEY, H. J. (2008), "Arab Identity and Ideology in Sudan: The Politics of Language, Ethnicity, and Race", *African Affairs*, Vol. 107, No. 426.

SHEARER, W. (1996), *The Victorians and the Race*, Aldershot, Scholar Press.

SIFNAIOU, E. (1996), *Lesvos. Oikonomiki kai Koinoniki Istoria. 1840-1912* (Lesvos. An Economic and Social History. 1840-1912), Athens, Trochalia.

SMITH, M. L. (1998), *Ionian vision: Greece in Asia Minor, 1919-1922*, London, C. Hurst & Co.

SNEL, E., STOCK, F. (2006), *Debating Cultural Differences: Ayaan Hirsi on Islam, Women and The Family*, Rotterdam.

SNOWDEN, F. M. (1970), *Blacks in Antiquity: Ethiopians in the Greco-Roman Experience*, Cambridge, MA, Belknap Press of Harvard University Press.

SPIVAK, G. C. (1988). "Can the Subaltern Speak?", in CARY, N. and GROSSBERG, L. (eds.), *Marxism and the Interpretation of Culture*, Urbana, ILL, University of Illinois.

SPIVAK, G.C. (1985), "The Rani of Sirmur", in BARKER, F. (ed.), *Europe and Its Others*, Colchester, University of Essex Press.

SPIVAK, G.C. (2003), "Righting Wrongs", in OWEN, N. (ed.), *Human Rights and Human Wrongs*, Oxford, Oxford University Press.

STANNARD, D.E. (1992), *American Holocaust*, Oxford, Oxford University Press.

STEFANI, G. (2008), *Colonia per maschi*, Verona, Ombre Corte.

STEPAN, N. (1982), *The Idea of Race in Science. Great Britain 1800-1960*, London, MacMillan Press.

STIANSEN, E. and KEVANE, M. (1998), *Kordofan Invaded: peripheral incorporation and social transformation in Islamic Africa*, Leiden, Koninklijke Brill.

STRATTON, J. (2007), "Dying to Come to Australia: Asylum Seekers, Tourists and Death", in PERERA, S. (ed.), *Our Patch*, Perth, Network Books.

STRATTON, J. (2009), "Uncertain Lives: Migration, the Border and Neoliberalism in Australia", *Social Identities*, Vol. 15, No. 5.

STRYKER, S. (2006), "(De)Subjugated Knowledges: An Introduction to Transgender Studies", in STRYKER, S. and WHITTLE, S. (eds.), *The Transgender Studies Reader*, New York and London, Routledge.

SURENDRA, L. (2004), "Indian Elections 2004: A Retrospective Analysis and Overview", *Kasarinlan: Phillipine Journal of Third World Studies*, Vol. 19, No. 2.

TAGUIEFF, P.A. (1987), *The Force of Prejudice. On Racism and Its Doubles*, Minneapolis, University of Minnesota Press.

TAKANA, Y. (2008), "The Politics of Local Boundaries and Conflict in Sudan: The South Darfur Case", Bergen: Chr. Michelsen Institute, retrieved from http://www.cmi.no/sudan/doc/?id=967.

TAMBINI, D. (2001), *Nationalism in Italian Politics: the Stories of the Northern League, 1980-2000*, London, Routledge.

TELEGLOBAUSTRALIA (2008), Intervista a Luigi Casagrande e Maria Teresa Restifa, candidati certi (alla data dell'intervista 24/02.2008 N.D.R.) per il PDL nella 4 Circoscrizione Estero, per le prossime Politiche 2008 in Italia, http://www.youtube.com/watch?v=zNzDG4JQtRg7.

TETI, V. (1993), *La razza maledetta. Origini del pregiudizio antimeridionale*, Roma, Manifestolibri.

THE INDEPENDENT (2001), "Fishermen hauled up corpses from 'phantom wreck,'" 8 June 2001, http://www.independent.co.uk/news/world/europe/fishermen-hauled-up-from-phantom-wreck-673190.html.

TIMES OF INDIA (2009), "Sonia Gandhi rules out becoming PM", 18 May, http://timesofindia.indiatimes.com/articleshow/683744.cms.

TINTORI, G. (2006), "Cittadinanza e politiche di emigrazione nell'Italia liberale e fascista. Un approfondimento storico", in ZINCONE, G. (ed.), *Familismo legale. Come (non) diventare italiani*, Bari, Editori Laterza.

TODARO RESTIFA, T. (2008), PDL Lower House candidate for the Italian political election in the Africa, Asia, Oceania and Antarctica district, speech delivered at the Marconi Club, Bosley Park, 28 March.

TORNARI, G. (2008), "Affrontare la realtà in un ottica internazionale. Gli italiani all'estero un ricco patrimonio di professionalità, conoscenza e talento da valorizzare per la madre patria", *La Fiamma*, 20 March.

TOSCO, A. (2002), "The Italo-Australian Press: Media and Mass Communication in the Emigration World 1900-1940", PhD Diss. School of Humanities, Faculty of Arts, Griffith University.

TOTARO, P. (2009), "The Lampedusa solution: Italy hardens heart to refugees", *The Sydney Morning Herald*, 7-8 February.

TRIPADVISOR (2004), "Quaint with nature: Christmas Island Resort", September 13, 2004, http://www.tripadvisor.com/ShowUserReviews-g295107-d459797-r2526332-Christmas_Island_Resort_Christmas.html.

TRIPADVISOR (2009), "La Calandra, Lampedusa: Traveller Reviews", http://www.tripadvisor.com/ShowUserReviews-g262047-d673698-r26667457-La-Calandra-Lampedusa.html#REVIEWS.

TRUSSEL, S. (1988), "The History and People of Christmas Island", http://www.trussel.com/kir/xmasi.htm.

TZOUMACAS, D. (2003), *Taksiki Sinidisi kai Eksidanikefsi sti Zoi kai sto Ergo tou Alekou Douka* (Class Consciousness and Idealisation in the Life and Work of Alekos Doukas), Melbourne, National Centre for Hellenic Studies and Research, La Trobe University.

UNCHR. (2009), "UNHCR worried about conditions for 2,000 boat people on Italian Island, UNHCR News Stories", 23 January 2009, http://www.unhcr.org/cgi-bin/texts/vtx/print?tbl=NEWS&id=497991064.html.

UNHCR (2009), "Irregular Migration by Sea: Frequently Asked Questions. Asylum in the EU, 28 May", United Nations High Commissioner for Refugees, http://www.unhcr.org/cgi-bin/texis/vtx/search?page=search&docid=4a1e48f66&query=pugh%20michael.

UNITED FOR INTERCULTURAL ACTION (2009), "List of 11,105 documented deaths through Fortress Europe", http://www.unitedagainstracism.org/.

VAN DER PLOEG, I. (1999), "The illegal body: 'Eurodac' and the politics of biometric identification", *Ethics and Information Technology*, Vol. 1.

VAN DER VEER, P. (2001), *Imperial Encounters: Religion and Modernity in India and Britain*, Princeton and Oxford, Princeton University Press.

VERACINI, L. (2006), "Settler Colonialism and Decolonisation", *Borderlads*, Vol. 6, No. 2.

VERDICCHIO, P. (1997), *Bound by Distance: Rethinking Nationalism Through the Italian Diaspora*, Madison, Fairleigh Dickinson University Press.

VERDICCHIO, P. (2006), "Horizontal Languages and Insurgent Cultural Alignments: National Popular Culture and Nationalism", http://lightzoo.blogspot.com/2006_03_01_archive.html.

VERDICCHIO, P. (trans.) (2006), *Translation of Antonio Gramsci's "The Southern Question"*, Toronto, Guernica Editions.

VERGOPOULOS, K. (1978), *Ethnismos kai Oikonomiki Anaptiksi: I Ellada sto Mesopolemo* (Nationism and Economic Development: Greece in the interwar years), Athens, Eksantas.

VILLA, R. (1985), *Il deviante e i suoi segni. Lombroso e la nascita dell'antropologia criminale*, Milano, Franco Angeli.

VINCENT, A. (2002), *Nationalism and Particularity*, Cambridge, Cambridge University Press.

VOUTIRA, E. (1997), "Population Transfers and Resettlement Policies in Interwar Europe: The Case of Asia Minor Refugees in Macedonia from an International and National Perspective", in MACKRIDGE, P. and YANNAKAKIS, E. (eds.), *Ourselves and Others. The Development of a Greek Macedonian Cultural Identity Since 1912*, Oxford New York, Berg.

WADIWEL, D. (2007), "'A Particularly Governmental Form of Warfare': Palm Island and Australian Sovereignty", In PERERA, S. (ed.), *Our Patch*, Perth, Network Books.

WALLERSTEIN, I. and BALIBAR, E. (1988), *Race, Nation, Class. Ambiguous Identities*, London, Verso.

WALLERSTEIN, I. and BALIBAR, E. (1991), *Race, Nation, Class: Ambiguous Identities*, London, Verso.

WALLEY, C. (1997), "Searching for Voices: Feminism, Anthropology, and the Global Debate over Female Genital Operations", *Cultural Anthropology*, Vol. 12, No. 3.

WARBURG, G. (1971), *The Sudan under Wingate: administration in the Anglo-Egyptian Sudan, 1899-1916*, London, Routledge.

WATSON, I. (2002), "Aboriginal Laws and the Sovereignty of *Terra Nullius*", *Borderlands*, Vol. 1, No. 2.

WATSON, I. (2002), "Buried Alive", *Law and Critique*, Vol. 13.

WATSON, I. (2007), "Settled and unsettled spaces. Are we free to roam?", in MORETON-ROBINSON, A. (ed.), *Sovereign Subject*, Crows Nest (NSW), Allen & Unwin.

WAUGH, L. (2005), "'Resort' as both noun and verb", *The Sydney Morning Herald*, 22 June.

WEEDON, C. (1987), *Feminist Practice and Poststructuralist Theory*, New York, Basil Blackwell.

WEST, C. (2002), "A Genealogy of Modern Racism", in ESSED, P. and GOLDBERG, D. T. (eds.), *Race, Critical Theories*, Malden, MA, Blackwell Publishing.

WHEELER, R. (2001), *The Complexion of Race: Categories of Difference in Eighteenth Century British Culture*, Pennsylvania, University of Pennsylvania Press.

WHITE, H. (1987), *The Content of the Form. Narrative Discourse and Historical Representation*, Baltimore, John Hopkins University Press.

WIEVIORKA, M. (1991), *The Arena of Racism*, New York, Sage.

WOLFE, P. (1994), "Nation and miscegenation. Discursive continuity in post-Mabo era", *Social Analysis*, Vol. 36.

WOLFE, P. (1999), *Settler Colonialism and the Transformation of Anthropology. The Politics and Poetics of an Ethnographic Event*, London, Cassel.

WOOD, C. A. (2001), "Authorizing Gender and Development: Third World Women, Native Informants, and Speaking Nearby", *Views from South*, Vol. 2, No. 3.

WOOD, E. M. and WOOD, N. (1978), *Class Ideology and Ancient Political Theory. Socrates, Plato, and Aristotle in Social Context*, Oxford, Basil Blackwell.

WOODWARD, P. (1985), "In the Footsteps of Gordon: The Sudan Government and the Rise of Sayyid Sir Abd al-Rahman al-Mahdi, 1915-1935", *African Affairs*, Vol. 84, No. 334.

WOODWARD, P. (1990), *Sudan, 1898-1989: The Unstable State*, London, Lester Crook Academic Publishing.

WYNDHAM, D. (2003), *Eugenics in Australia. Striving for National Fitness*, London, The Galton Institute.

YEROLYMPOS, A. (2003), "Inter-war Town Planning and the Refugee Problem in Greece: Temporary 'Solutions' and Long-Term Dysfunctions", in HIRSCHON, R. (ed.), *Crossing the Aegean: An Appraisal of the 1923 Compulsory Population Exchange Between Greece and Turkey*, New York, Berghahn Books.

YIANNAKOPOULOS, G. A. (1992), *Prosfigiki Ellada* (Refugee Greece), Athens, A.G. Leventi-Kentro Mikrasiatikon Spoudon.

YOUNG, R. (1995), *Colonial Desire. Hybridity in Theory, Culture and Race*, London & New York, Routledge.

ZINCONE, G. (ed.) (2006), *Familismo legale. Come (non) diventare italiani*, Bari, Editori Laterza.

ZIZEK, S. (1997), "Multiculturalism, or, the cultural logic of multinational capitalism", *New Left Review*, Vol. I, No. 225.

MIX
Papier | Fördert
gute Waldnutzung
FSC® C083411

Kolibri 360 GmbH
Ferdinand-Jühlke-Straße 7
99095 Erfurt, Deutschland
produktsicherheit@kolibri360.de